Constructing Canine Consent

The concept of canine consent is far more than simply a buzzword in modern dog training practices. In its current form, consent is a distinctly human concept, designed by humans and for humans. Looking beyond species boundaries can help us not only consider concepts of canine consent and autonomy, but it can also help us to apply these concepts in our everyday interactions with dogs, which is fundamental for any professional working with dogs as well as for everyday dog caregivers. This canine-indexed definition of consent includes a model of five major categories: Touch/interaction-based consent, cooperative care using learned consent behaviours, activity consent, consent-based learning, and substitutive consent. These categories involve a two-way communication system, integration of salient choices, teaching consent behaviours and incorporating existing training protocols that adhere to the Humane Hierarchy of best practices, and an evaluation of dependent decision-making in extenuating circumstances. This book aims to merge the existing literature and new understandings about canine consent to paint a complete picture. It will challenge the current expectations of dogs and dog behaviour in our society with an intention of considering their perspectives, experiences, and emotional needs. It will be important reading for veterinary professionals, dog trainers and behaviourists, those involved in work with therapy dogs, and anybody working with or caring for dogs.

Constructing Canine Consent

Conceptualising and Adopting a Consent-focused Relationship with Dogs

ERIN JONES, PHD, CPDT-KA, IAABC-ADT & CDBC

CRC Press
Taylor & Francis Group
Boca Raton London

CRC Press is an imprint of the
Taylor & Francis Group, an **informa** business

Designed cover image: Erin Jones

First edition published 2024
by CRC Press
2385 NW Executive Center Drive, Suite 320, Boca Raton FL 33431

and by CRC Press
4 Park Square, Milton Park, Abingdon, Oxon, OX14 4RN

CRC Press is an imprint of Taylor & Francis Group, LLC

©2024 Erin Jones

ISBN: 9781032421674 (hbk)
ISBN: 9781032421599 (pbk)
ISBN: 9781003361459 (ebk)

DOI: 10.1201/9781003361459

Typeset in Janson
by Apex CoVantage, LLC

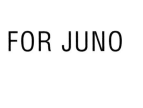

CONTENTS

Juno dips her head into her harness and navigates her way to the front door. She anticipates the sweet smells of the world, the earth, the dew, the leaves, and bugs. But also, the urine, garbage, and bird faeces. She stops at the bush at the end of the driveway, one nostril inhaling air and then the other. Her eyes soften, and she takes a moment to process who has passed by her house recently—who was there, when, and for how long. The sun is just starting to peek over the horizon, and she squints as the sun flares obstruct her view. It doesn't matter. She can navigate perfectly well without her sight fully operational. She walks slowly, then quickly, and then side to side. She stops and starts. She jumps up on the benches and scavenges for food wrappers. She frequently pokes my leg with her nose whenever we see a person, and I toss her a treat and smile; I tell her how brilliant she is. We walk together, friends, enjoying each other's company. Taking our time to watch the birds, to smell the wafting scent of warm muffins in the café, to have small passing visual conversations with other dogs. Our outings are for her, for us. They are not for me alone, not for the child who wants to place his hands on her to stroke her soft fur, not for the people who smile and tell me how pretty she is, and not for the elderly woman who stops to ask about how old she is or what her name is. This time is for her to exist without anyone invading her space, without the threat of unsolicited touching. I will never ask her to disengage with smells or not look at the dog across the road. I will never shorten the lead and force her to move in a continuously straight line at a continuously human pace. It doesn't mean there are no guidelines, and it doesn't mean she has entirely free will to do as she pleases. We are sharing our space with others who also matter, after all. That's why I have provided her with skills to feel safe and to be successful around others. And so we enjoy our time and wind our way back along the river and to our home. I sit down and begin to write, and she burrows into a blanket on the couch, head resting on a pillow, and we both feel contented.

Juno is my canine companion; she lives in my home, and we share our experiences respectfully. Juno will make several appearances throughout this book to guide you through examples of consent, autonomy, choice, and preferences. In fact, my most superb teachers in life have been dogs. Yes, I have learned from some of the great behaviourists about learning science and animal behaviour. I also have an honours bachelor's degree, master's, post graduate diploma and PhD, but no one human supervisor, professor, or lecturer has ever rivaled the education that dogs have gifted me. I suppose these profound teachers have mostly been the dogs who have shared my home. However, I have worked with thousands of dogs in my career as a behaviour consultant and trainer, and they, too, have taught me everything from the importance of collecting data, to observational skills, to how to enter into a cross-species relationship effectively and ethically, including how consent is and should be applied.

Consent is perhaps a novel and somewhat contested concept in the discourse of our relationship to animal others, but I argue, if we can shed our anthropocentric bias and redefine consent in a way that makes sense to dogs and for dogs, there is no reason to deny dogs the right to consent in most situations. When applied to nonhuman animals, the word "consent" itself is often controversial, but it is a term I use very deliberately to highlight that it can and should be used to decentre the common paradigm of **human exceptionalism**. That is, dismantling the ascendancies left behind from traditional antagonists of animal liberation theories to engender a praxis for dogs as consenting individuals within a shared society.

The seedlings for this book were planted when I was working as a behaviour consultant in a veterinary clinic and made the decision to go to graduate school. My intention was to start a master's degree in applied animal behaviour, but I came across a fascinating programme in anthrozoology. This was the first time I had heard of this field, and it piqued my interest immediately. I wasn't even sure why; it was more of a feeling than a checkmark on any specific life agenda. When I met my cohort, I knew I had found my home. From there, my interests in critical animal studies grew, but I really wanted to be able to merge my work as a certified dog behaviour consultant (and trainer) with critical theories of animal rights. This felt as if it were an impossible task—the fields are virtually polarised in many ways, from the language and methodological tools they employ to the intention behind the design, hypothesis, and findings of their research. But I don't necessarily see it that way; I actually see the advantage and superbly powerful pedagogy and praxis that both sciences provide when used in complementary ways. Behaviourism, psychology, and ethology are all fields that bring measurable and observable evidence that allows us a glimpse into the minds of other animals by demonstrating their abilities, strengths, and sometimes comparable differences from humans. Critical animal studies bring an interdisciplinary methodology that melds activism to theory (and theory to activism). Though it may not be easy to merge the literature between these disciplines seamlessly, that was the aim of my PhD thesis—to highlight critical theory and qualitative findings with quantitative studies of behaviour in its operational forms. And it was this writing and research process that helped me immensely with the process of structuring this book. The contents are rooted in that epistemological cross-disciplinary approach, converging the cognitive behavioural sciences to enhance critical theory in a way that further highlights exactly why scholarly work needs to focus on difficult topics like nonhuman animal consent—and that scholarly work needs to have a practical application for practitioners who use it to effectively shift the current social paradigm to one that includes animal others in more ethical ways. Theory only matters if we can use it in practice to make effective change, and my dog teachers have helped me to utilise my knowledge in practical settings.

Most of my past dog teachers were social, well-adjusted, and resilient, but Juno is the exception. This is what makes her a perfect archetype for why canine consent matters (only because it's more obvious, not because her feelings matter more). She is a sensitive soul with a lot of anxiety. And she is a real free thinker in many ways. I have always supported her ability to advocate for herself and voice her opinions and consent (or not). It is Juno who ultimately inspired my research into the connections between human exceptionalism and the paternalistic relationships we have with dogs—the "caring-controlling paradox" that sends most of us into a dither. The relationship is one of both love and dominion, a Cartesian dualism demarcated as human and nonhuman, or human apart from nature, that lingers on in our relationship with other animals. Perhaps, as this book will uncover, humans are well-positioned to be both the foreseer of "a good life," making decisions that improve the lives of our dog friends, as well as the ultimate creator of these disempowered dependents, making them ours in the first place, rather than empowered agents of their own being. Relationships are forged, but at what cost, and for what purpose?

This book is about how we can create a better life for our dogs and engender a better understanding of their experiences. I don't pretend to have all the answers. Science is forever allowing our knowledge to grow and expand our awareness. And when we know better, we do better; so our better is forever getting better (hopefully). What I do know is the trajectory of the dog–human relationship needs to be one that is consent-based and respectful of bodily autonomy and empowers our canine friends to be thinking and skillful adults, allowing them to thrive in a shared society that embraces our differences from others and respects their rights.

ACKNOWLEDGEMENTS

First off, I want to thank the dogs in my life who have been my greatest teachers over the years: Willie, Cassie, Zoe, Jake, Charlie, Monday, Stewart, Blue, Layla, Maggie Mae, Iggy, and Juno. Without them this book would have never been inspired or composed. On the human side, there has been some tremendous help from colleagues, family, and friends who offered ideas and concepts and contributed to my vision. Thank you to Jessica Benoit, Emily Tronetti, Patrick Flynn, Mia Cobb, Jade Fountain, Lavinia Tan, Clara Fox, Lisa Sturm, and Maria Alomajan for your help with the development of this book in various ways and through countless conversations about this topic over the years. Thank you to Nik Taylor, the most extraordinary PhD supervisor and colleague I could ever have the chance to work under and alongside. To Jessica Pierce who not only deeply inspired my work with her own, but who, as my primary PhD examiner, gave me the courage and support I needed to undertake this project. Thank you to my supportive husband, Mike Thompson, for always being my biggest ally. Thank you to my sister, Amy, who will always be a much better writer than I am but motivated me to aim higher. Thank you to my parents, Shelagh Hagen and Richard Jones, and to my late mother, Bonita Jones, who would have been my sounding board if she were here.

Erin Jones is a scholar, author, and applied behaviour consultant. Her master's and doctorate research are specifically rooted in critical animal studies. She combines her applied practice as a certified dog trainer and certified dog behaviour consultant with a deeper academic background in how society treats dogs and the entanglement and complexities of multispecies relationships. She has a BSc. in psychology and anthropology from Trent University, Canada; an MSc. in anthrozoology from Canisius University, USA; a PhD in human-animal studies from University of Canterbury–*Te Whare Wānanga o Waitaha* (at the New Zealand Centre for Human-Animal Studies); and a Pg.Dip. in animal welfare from Thompson Rivers University in Canada. She also is a certified professional dog trainer (CPDT-KA) with the Certification Counsel for Professional Dog Trainers, an accredited dog trainer (ADT), and certified dog behaviour consultant (CDBC) with the International Association of Animal Behaviour Consultants. She also works independently and contractually in several sectors of animal studies research, including topics beyond the dog–human relationship, everything from ethical veganism to topics in critical animal studies. Erin operates Merit Dog Project, an educational platform used to educate and inspire industry professionals and front-line specialists working primarily with dogs/dog behaviour. She currently lives in Christchurch, New Zealand.

CHAPTER 1

CONSTRUCTING CANINE CONSENT

"Our ideas of consent have evolved and changed because feminism has pushed the boundaries."

—Barkha Dutt

1.1 INTRODUCTION

Over my lifetime I have done some casual travelling throughout South and Central America and was introduced to a dog culture very different from what I'd grown up with in Canada or where I currently reside in New Zealand (See Figures 1.1 and 1.2 for two individuals I met in Costa Rica and Ecuador). Most strikingly, there were free-roaming dogs sleeping on beaches and choosing to sit beneath tables at restaurants and wandering through the streets, unattached and living liberally. In fact, it is estimated that over 70–80% of dogs globally are free roaming/free living (Gompper, 2014; Hughes & MacDonald, 2013; Lord et al., 2013). Some of these dogs I met along my travels likely had homes to return to, others maybe didn't, but they all had some form of autonomy.

Free-roaming dogs seem to be culturally normalised in many countries, and in my passive observation, the dogs mostly seemed relaxed, or at least well habituated to their environment, which is demonstrated recurrently in the book *Dog Knows: Learning How to Learn from Dogs* (Pangal, 2021). Sindhoor Pangal (from the organisation BHARCS, a research and education institute in Bangalore, India) suggests that free-living dogs are often calm; they are easily able to regulate their emotions and rarely engage in "frenzied activities" (BHARCS, 2023).

While working in a very remote area of Pacuare, Costa Rica, one summer, a local dog frequented our area of the beach where we nested sea turtle eggs for hatch and release. A young American girl with whom I was working, someone with the best intentions mind you, wanted to "rescue" this particular dog. She began looking into ways to export her back to New York. Although this dog could likely benefit from vaccines and deworming treatment, she really didn't need "rescuing." She visited who she liked when she liked. She was clearly well fed, social, had local canine and human friends, and benefitted from the freedoms of being unowned (or at least unmanaged). But most of all, taking her from this lifestyle seemed cruel and

DOI: 10.1201/9781003361459-1

unnecessary. The adjustment to city life would likely be unbearable for her. Who would benefit from this "rescue," this dog or the person doing the "rescuing?"

In reality, it wasn't that long ago that companion dogs in Canada had more independence too. In the 1980s (though things were starting to change), my mother always let our dogs out the front door for a wander despite having a fully fenced yard. It wasn't uncommon; lots of people in our neighbourhood did the same. In fact, where I lived when I started writing this book, in Tauranga, New Zealand, it also wasn't exceptional to see dogs freely roaming the parks and streets in my neighbourhood without allegorical or literal human ties. I don't know if this is intentional or not. However, what strikes me is they (generally) are not displaying signs of stress or hyperactivity, nor do they seem worried about much of the happenings in their environment, and only very occasionally does one behave "aggressively" (for all intents and purposes). Granted, those dogs with a tendency to react unfavourably are probably less likely to be granted wandering privileges, but these roaming canines also don't behave the way dogs living in mostly solitary, confined conditions do when they finally get their 30 minutes of "freedom." It makes me wonder: How much do our increasing restrictions influence the behaviour issues we are seeing on the rise? There is certainly an abundance of variables that contribute to overly reactive or aggressive behaviours, but I highly suspect that micromanagement, society's restrictions, and a suppression of a dog's autonomy are amongst them. This is principally why it is important for us to consider consent and how it applies to our companion dogs in various ways, including all of the other considerations that accompany

Figure 1.1 Free-roaming dog under my table in a restaurant in La Fortuna, Costa Rica.

Figure 1.2 Free-roaming dog near Otavalo, Ecuador.

consent: Choice, agency, preferences, and cooperation are all factors that play a starring role as we move to a more progressive relationship with dogs.

1.2 CONSIDERING ANIMAL OTHERS AND WHY CONSENT MATTERS

I ask that you open your mind and heart as we dive into this important but challenging topic of why consent matters. Drawing from multiple disciplines, we can better understand the role of dogs in society and how that influences the way we treat them. Thus, although I include and critique animal rights and welfarist theories, and the humbling influence of human exceptionalism,[1] this is truly a discussion of ethics and the moral inclusion of animal others in our world (and us in theirs). Ethics is important in understanding the depths of our entanglements with dogs; it is the foundation of why consent should matter. The only way forward is to understand why we struggle with concepts of canine consent and to challenge lessons from the past to create a more cooperative and trusting relationship with the very important dogs in our lives now and in the future. The way we treat other animals rests largely on the way we have constructed them to fit into our lives, but it is important to also look at the ways that we fit into (or even disrupt) theirs.

Dogs play many roles within society. They may be perceived as surrogate children, companions, workers or service providers, athletes, entertainment, experimental subjects, or partners (Cunningham-Smith & Emery, 2020; Serpell, 2017). These multiple roles have led to different, often entrenched, positions about any rights that dogs may have. The discussion of animal rights theory (ART) and other alterative theories can bring some depth to this discussion if we are to think about dogs as individuals who both are able to, and have the right to, consent.

Some animal rights scholars believe that keeping dogs as pets is in fact enslavement, that dogs are the product of human ascendancy, and that dogs should either be returned to their "wild status" or be extinguished entirely from existence (see, for example, Francione, 2000). It is traditional ART that advocates liberty for other animals but remains dubious of a just and egalitarian mutual relationship between humans and other animals. They argue that we should "let animals be," separate from human society, and cut our ties with those reliant on humans (Palmer, 2010). This is not representative of all ART, as we will see, but it is this radical, total liberationist view that prevents many people from engaging with other valuable constructs in ART that could benefit our relationships with animal others.

Alternatively, welfarist approaches to animal ethics start from the premise that continued relations between humans and other animals are inevitable, and therefore, reformation of these relationships is our ethical duty. It sounds great, but welfare theories more or less suggest a demarcated characterisation of other animals' interests. This is highlighted by examples that emphasise their suffering while ignoring their independence or by idealising the human–animal relationship as one founded on love and loyalty while ignoring the tyrannical dimensions. Even the most generous welfarist theories often deny that animals have a vital interest in living autonomously, let alone considering the concept of nonhuman animal consent. Let's face it, how many animals would likely consent to the way we use and exploit them for our benefit in the various ways we do, such as the most obvious examples of factory farming, circuses, or laboratory testing, but also being tethered and commanded and prevented from behaving in ways that foster feelings of self confidence and empowerment. Even if we are committed to reducing their pain and suffering while in our care, it is aimed at how we can reduce their pain and suffering so we can benefit from their bodies in various ways (as test subjects or food or for the materials they produce for humans to use).

ART and welfarist theories offer some great context and information but none offer a wholistic account of how we can maintain socially meaningful cooperative relations with dogs while still protecting their fundamental rights. There are some scholars who see a way forward that is more inclusive of dogs as citizens in a shared, less exploitative coexistence (see, for example, Donaldson & Kymlicka, 2011, which we visit in more detail further on). There are potential ways we can be more inclusive and more accommodating of other animals with whom we share our world and our homes. This includes altering our own expectations to better appreciate dog culture as unique from our own and incorporating a dog-specific consent to our interactions.

1.2.1   Animal Rights and Human Exceptionalism

To challenge our existing expectations and intentions about dogs, it is important to examine the deep-seated influence of human exceptionalism. Human exceptionalism is a paradigm that humans are different, better, and above all other animals. It's partly a derivative of traditional Cartesian dualism, theories that were fundamental to Enlightenment rationalism and partly a derivative of Christianity.

Within Christian tradition, there is a narrative that is stranglehold to the uniqueness of humanness: Humans are inimitable beings—they are made in the form of God, and they have a rational soul (Peterson, 2001). John Gray (2016) suggests that persistent human exceptionalism is "a secular religion thrown together from decaying scraps of Christian myth," tarnished beliefs that seep into our everyday interactions with animal others. Peterson (2001) writes that within Christianity it is believed that "the soul is not just another piece of equipment but a singular dimension that transforms the meaning of humanness" (p. 29); the fact that we cannot measurably observe the soul has led many secular thinkers to reject the concept of the soul altogether. However, it is this inability to quantify the soul that can also support assertions of human exceptionalism. For example, Peterson says, "while evidence might show that chimpanzees can learn a language, there is no way to prove that one might have a soul" (p. 29).

Rene Descartes also considered the mind to be distant from the body and that mental consciousness is the heart and lifeblood of human nature. Thus, according to Descartes, humans are unique in that they have a mind, the core of humanness, and other animals do not. Because Cartesian thinking supported a denial of other animal minds, it also meant that other animals could not have personhood, moral worth, or rights. Anderson and Perrin (2018) claim that the idea of human exceptionalism can

> be traced back from recent conceptions of culture as a distinct realm of human agency, to Descartes's identification of the human being's unique "mental" capacities with an immaterial mind, and then further back still to Christian ideas about the soul.
>
> (p. 449)

In recent years, ART, cognitive behavioural studies, and philosophy have all challenged (and defeated) the suggestion that animals don't have awareness, a sense of self, or even personhood (Rowlands, 2019); and certainly there are not many people who would argue their dogs don't have a mind. However, this dualistic thinking (the human and nonhuman animal divide) has shaped the way we perceive and treat other animals despite our recent advances in knowledge.

Because of these institutionalised conceptions that have largely shaped the framework of modern relations with nature, human exceptionalism in today's social climate remains systemically embedded, a socially rooted structure that parallels other human oppressions. Hierarchal and often binary social constructs—like the human–animal divide, race, class, gender, disability, and more—work to oppress groups based on characteristics that are given more or less value. Companion dogs are discursively constructed in ways that legitimate our dominative and paternalistic role, exemplifying the inherent asymmetry existing in our relationships with them and their "use" to humans (Jones, 2022; Pierce, 2016; Stibbe, 2001). This favours a representation of non-human animals that support an anthropocentric hierarchy, rather than a "horizontal position" that considers more equitable structures (Marchesini & Celentano, 2022).

Although companion dogs are fundamentally dependent upon humans throughout their entire lives, there is the potential for a good, shared life that isn't exploitative and considers their agency, even when their agency is effectively dependent on macro decisions made on their behalf. Fundamentally, ART begins from the assumption that every living creature with a unique, subjective point of view on the world matters morally in the same basic way. Anyone with a life that can go well or poorly, who has the capacity for hope, desire, or disappointment—anyone with the traits that we see in our dogs every day, that make them such beloved companions—should expect to be considered as part of "who matters." If we care about doing what's right, about acting in a way that promotes justice and well-being and steers us away from cruelty and suffering, then we have no good reason to exclude dogs from our moral position. Learning to listen and to respect consent wherever it can be given is good for the individuals who are expressing themselves, and it's good for their caregivers and for society as a whole.

It is important to make clear that believing dogs have rights does not mean believing dogs should necessarily be treated the same as humans, that anyone (canine or other) is always entitled to free choice, or that every interaction between a human and a dog is morally suspect. Many dog experts and caregivers express concern that ART creates an extreme opposition, suggesting that dogs are living miserable lives, comparing them to enslaved humans, or even determining that dogs as a species should not be allowed to exist at all (Greenebaum, 2009). As I have mentioned, this extreme abolitionist view (which is not representative of the spectrum of views of ART scholars) holds that there is no ethical way for humans and dogs to live together. This view is understandably horrifying to anyone who lives with and loves dogs, myself included. However, it is possible to be committed to animal rights without disengaging completely from relationships with dogs. Indeed, as Barbara Smuts (2001) puts it, we feel the presence of another self in other animals more profoundly than simply knowing they are there.

The problem of other animal minds, however, is that humans can only infer phenomena from physiological and ethological responses and behaviours (which is really a human problem, actually). Comparing dogs' capabilities with humans' capabilities only highlights the need for a modified dog-index of consent and for a modified paradigm about how we view companion dogs in general. For years, ART scholars have suggested that regardless of whether other animals function in the same way humans do, they are deserving of equal consideration in their own right (Singer, 1973). Underpinnings of human exceptionalism has, for far too long, created a rift between "animal" and "human"—a false dichotomy influenced in many directions and solidified by the social construction of the human as transcendent above nature, when in fact we are simply one small part of it. Philosopher Lori Gruen (2021) faults any human-centred approach as being speciesist and discriminatory and thus calls for what she has coined as "entangled empathy," a process of affect and cognition that goes beyond an empathetic approach to relating to other animals, and enhancing critical evaluation of their realities.

ART scholar Tom Regan (2004) argues that rights should be based on the intrinsic value of consciousness. Once we concede this intrinsic value that animals and humans share, animals should also have the same moral rights as humans. For Regan, then, the injustice is not only in animals suffering pain from human exploitation, but in viewing other animals as our resources. The upshot is that we should abandon treating animals as instruments for human ends. However, while this argument is often applied to animals used for science or those commodified for various industries (like using their bodies for food and material goods), our "pets" lie in a precarious middle ground between subjugation and incontrovertible human care and concern. But care on concern doesn't negate injustices.

There is an ambiguous conceptualisation of dogs as both inimitable individuals and disposable resources who can be the subjects of breeding, buying, selling, and the like (Redmalm, 2019). Dog ownership can be viewed as a delineated sphere where biopolitical values about rights and life can be engineered in ways that suit our own conveniences. It comes in the guise of affection, but perhaps, as Tuan (1984) suggests, affection is inseparable from dominance; making a dog perpetually reliant on us is, in fact, a form of dominance. We have dogs for the affection and "companionship" they provide, but these ideas are partly shaped by our expectations marked by an underlying human exceptionalist view. For Tuan (1984), creating pets is an exercise in power that requires the dog to conform to and comply with specific behavioural criteria to obtain affection and support from their "master."

It was in the nineteenth-century that pet-making in Western society involved displacing affection away from human family members and toward pets (Keralis, 2012). According to Tuan (1984), this is when human affection became increasingly challenging to formulate as "modern society . . . began to segment and isolate people into their private spheres" (1984, p. 112). Though, largely, companion dogs have not provided services in the household, they fulfil(ed) an aesthetic and emotional need for their caregivers. The sentimental aestheticisation of dogs through practices of selective breeding has produced an array of morphologies—from miniature breeds that fit into handbags to brachiocephalic pugs and bulldogs, some of which are unable to give birth naturally or breathe effectively. The operation of power itself has been embodied in these dogs through these practices of dominion. That is not to say that we have not also bred for function in certain breeds—for example, services such as protection and hunting—but we then attempt to quell the behavioural consequences (barking, chasing, etc.) as these are reconstructed as "undesirable" when out of context of their original intended use.

There have been obvious objections to the reduction of our relationship with dogs as mere dominance and paternalism. A number of dog-loving people, myself included, who also appreciate the symbolism of having dogs in our lives, have contested Tuan's writing on the interconnectedness of dominance and affection. Erica Fudge, who defines Tuan's theory as "uncomfortable," suggests that "pet ownership is premised on the notion that it is possible to extend one's capacity to love beyond the limits of species; that one can have a truly affectionate and meaningful relationship with a being that is not human" (Fudge, 2014, p. 13). Tuan emphasises the importance of affection, care, and concern in the pet-human relationship but argues that these things veil the inherent violence of creating the "docile and friendly pet." According to Keralis (2012), there is an almost reflexive dismissal of Tuan's theory by many dog caregivers, though the form and the fervour of the dismissal seem to reinforce his point. So while we claim that pet keeping is built on care, the institution enacts forms of oppression that permeate our social organisation, upholding a nurturing yet highly paternalistic form of dominance that models the structures of the patriarchy itself and valorises a benevolent subjugation. We enact our love through oppression, in the discourse of servitude, in expectation of obedience and in the suppression of the will of our beloved captives.

So as we circle back to a traditional extremist ART approach of abolishment, I argue that we cannot simply abandon dogs whom we have created to be dependent upon us. Also, there is no evidence to suggest that some form of pleasure and reciprocity does not exist for our dogs. While some animal rights activists' claims about how dogs are treated are based on uncomfortable truths about oppression and a dominative framework, the abolitionist point of view is not the only conclusion we can draw from the premise that dogs are the sorts of beings who matter morally. In fact, giving dogs the respect that they deserve means bringing them *more completely* into society, not abandoning them. And by doing so we need to consider their right to agency, natural enactment of behaviour and communication, and a dog-centred approach to consensual interactions.

As briefly mentioned earlier in this chapter, Donaldson and Kymlicka (2011) have developed a political theory of animal rights that argues against abolitionism by working from a conception of relational rights and community membership. They argue that no matter what we think about the morality of domesticating a wild animal species—arguably as humans did to canids in the distant past—dogs now live as part of our communities. Donaldson and Kymlicka posit that while our duties to wild animals are best summed up as "let them live freely in their own communities," for our companion dogs, their community *is* our society. They depend on us, and we have a special set of obligations to them because we made them dependent animals that are expected to suppress many of their natural, instinctual behaviours and conform to human lifestyles.

Some of these concepts may be painful to contemplate—it is true, our dogs carry many values for us, to us, with us, and in spite of us. For many, dogs are much more than simply personal belongings, tools, or status symbols. They are also close companions with whom we share our lives and our love (Irvine, 2004). It would be imprudent to diminish this bond to a mere underlying need or want to dominate them. The relationships we have with companion dogs are significantly varied but often genuine. However, it should be noted that this is not always the situation, as we see a high rate of rehoming, abandonment (Coe et al., 2014; Manning, 2013), and abuse (Manning, 2013). Our proprietary role is also likely to serve as a way of reducing the position of "pet" in a relationship of power; "pets" are handled and viewed differently according to the arbitrariness of human ways (Eddy, 2003). An inherent problem with relationships across species boundaries means that people often struggle to accurately interpret and even empathise with their dogs' emotional realities. We often ascribe feelings and opinions to our dogs, often inaccurately, which ultimately influencing how we choose to interact with them. This is partially because most people are not well-educated on dog behaviour and also because our human lens, steeped in human exceptionalism, intrinsically biases our interpretations and fosters unrealistic expectations, as we will later explore. People are required to navigate the social and enforced rules, laws, and norms that govern "pet keeping" and it becomes a struggle for humans to try to relate to dogs in a more egalitarian way (Smith, 2003).

Thus far I have argued that human exceptionalism influences the way we socially construct dogs. These expectations create an asymmetrical and paternalistic relationship constructed in dominance despite the love and appreciation we also have for dogs. In part, paternalism is the result of the infantilisation of dogs, a process of both humanising dogs to be a part of the family but dehumanising them so that they remain less than human. In the remaining sections I more explicitly introduce concepts of consent, but I feel it is important to always keep in mind the influences of our position over, with, and to dogs and how we can aim to adjust that role in ways that focus on their particular experiences and interests rather than our own.

1.3   DEFINING CONSENT

Before we engage specifically in the topic of consent, what it is, and how to (re)define it for our canine friends, I think it is important to understand some related terms that all interconnect in necessary ways. These terms will be used throughout this book as we grapple with the tribulations of a canine-specific consent.

1.3.1   Autonomy and Agency

Being autonomous means having the ability to make decisions independent of external control, and agency is the power to act on that freedom. Free-roaming dogs have a lot of autonomy and agency, whereas companion or working dogs have very little, if any. Autonomy is connected to consent—the voluntary agreement for something to occur—in that it may afford dogs the freedom to choose when and how to interact and when to opt out. In a sense, it means that individuals can, to some degree, develop their own sense of value, make decisions about what matters most to them, and act accordingly in those situations. Studying free-roaming dogs is one quintessential way for us to better understand the importance of choice, autonomy, and agency because they have the freedom to choose to eat, sleep, and play when, where, and with whom they please, even if there are some external influential environmental factors. Therefore, I will frequently provide examples from the literature on free-roaming dogs throughout this book to highlight some key concepts. On the other hand, we regulate play, friends, food, sleep, space, access, resources, learning outcomes, opportunities, and just about everything in between for dogs that live in human care, leaving them little opportunity to have any true control over their own lives.

However, it's not quite that simple. As we will discover throughout this book, simply providing choices and offering (supervised) freedom doesn't actually equate to autonomy or consent in the way we may hope. For example, I may resolve to only use a leash to restrict my dog, Juno, in emergencies, but that doesn't mean that she has true agency the rest of the time. It means she has a **conditional agency**—conditional on my desire to grant her agency, which could essentially change at any moment for reasons she is powerless to change or understand. So companion dogs don't often have true autonomy, just a *feeling* of autonomy. But dogs, as a species, certainly have the ability to be autonomous, and they, like other animals, grow into fully functional adults. However, it's humans who often prevent a fully autonomously lived life by selectively breeding dogs to be perpetually dependent and by keeping them in ways that prevent their independent success.

Much debate occurs in academic circles whether other animals require metacognition—an awareness of their own thought process—in order to be autonomous. Some would say that autonomy isn't just about having a sense of self control, but it is a **justified belief** about that control. For example, Cochrane (2009, pp. 667–668) claims that since nonhuman animals do not have the capacity to reflect on their own desires or values, most other animals are in fact not autonomous. This stems from a long legacy of hierarchal thinking (for example, Cartesian dualistic theory), though there are some scholars taking the liberty of other animals seriously. Valéry Giroux, for instance, disputes previous philosophers' assumptions that other animals have no interest in living autonomously because they are not rational. She states that

> even if sentient nonhuman animals are moved only by their instincts or their desires and never by their reason, they are harmed whenever an external agent keeps them from behaving as they are inclined to behave, since—as is the case with humans—this causes them pain, fear, anxiety, or, at the very least, frustration. It seems reasonable to conclude that these animals have a certain type of intrinsic interest in liberty.
>
> (Giroux, 2016, p. 33)

In my earlier example about free-roaming dogs, their capability to reflect on or even value their own autonomy doesn't necessarily mean they can't and don't experience it. Gruen (2011, p. 150) argues that there are many autonomous animals that make their own choices about "what to do, when to do it, and who to do it with." Furthermore, simply because we don't have the knowledge or capacity to measure self-reflection in other animals doesn't automatically rule it out. Assumptions based on human bias are not good enough to justify the control we imbue on other animals. How (and even if) autonomy and agency can be achieved with our companion dogs is something this book will challenge in the coming chapters.

Clearly there is a convincing argument that dogs *can be* autonomous, though companion dogs are *not* fully autonomous because of their relations with humans. However, all of this is based on whether dogs are ever able to experience autonomy in the same way humans can. Does that even matter? In the discourse around consent, perhaps it doesn't matter in the way one might think. Autonomy is a tricky thing; as we move on from Enlightenment rationalism, we know that we don't need to be a certain person (example, a White man) be able to have our choices be respected. This demand for a kind of intellectual sovereignty of the individual has been informed as much by social and political power structures as it has by philosophical argument, and we're just about starting to get past it. Shifting this argument from human to nonhuman animal is moving beyond speciesist boundaries and can be a difficult concept to engage in, but I have no doubt we will, can, and are. We now understand that kids can consent, women can consent, disabled people can consent, and that people who think in ways other than the way Enlightenment rationalists do can consent. And we are just starting to expand the circle, discussing how animals can consent, separate from their ability to live autonomously. So although autonomy is important, we don't need a conception of autonomy to expand, we simply need a conception of the role consent plays in moral interactions.

We may have a long way to go, but we have to begin to conceptualise what this will look like for dogs, specifically, when we remove the human connotations. How (and even if) autonomy can be achieved with our companion dogs is something this book will challenge in the coming chapters.

Key Questions

1. Does withdrawing consent mean that any action that comes after the withdrawal is wrong?
2. Does an individual consenting to something mean it is automatically permissible to do it to them?
3. Is consent one part of the complex machinery of obligation, respect, desire, choice, freedom, etc. that produces right and wrong between people?

1.3.2 Choice

Being able to choose between two or more options is empowering and may be one way to increase at least the feeling of agency. Although true choices are far and few between for our companion and working dogs, a study by Perlmuter and Monty (1977) showed that simply having the illusion of choice can be invaluable to well-being. Ironically, some well-documented examples come from captive settings. For example, when zoo-housed giant pandas were given the choice to go into an enclosed space out of public view, they were less agitated and showed a decrease in stress even when they chose to remain outside (Owen et al., 2005). Similarly, Anderson et al. (2002) found that when goats and sheep in a petting zoo were given the option of a human-free "retreat space," they too showed lower rates of behaviours often associated with stress.

Even just the choice to participate or not in tests of an animal's cognitive capabilities can lead to positive outcomes. A study by Whitehouse et al. (2013) found that crested macaques living in zoo enclosures have demonstrated positive welfare benefits from choosing to voluntarily separate themselves from their group to participate in cognitive testing. Another example comes from Perdue et al. (2014) who found that rhesus monkeys who were asked to complete a sequence of four touchscreen-based tests preferred to choose the order in which they completed them rather than completing them in a random arrangement (Perdue et al., 2014). Likewise, having a choice to control lighting in their enclosures led laboratory-confined marmosets to show a significant increase in calm behaviours (Buchanan-Smith & Badihi, 2012).

It's unfortunate that such convincing evidence has emerged from testing captive animals. Captive-living animals clearly have reduced control over their lives compared with their wild counterparts, and the irony is not lost on me. Having said that, the research exists and has paved the way for our understanding of the importance of personal control and how we can provide choices to create the best possible environments for the animals in our care. Be mindful that these studies only better an arguably less than ideal situation for animals in captivity, but they do provide some tangible measures for our companion dogs as we move into a more consent-focused relationship.

In reality, our dogs don't have a lot of opportunity to make choices either, so it's vital that we create as many options as possible. Perhaps one of the most difficult parts of providing choice (and also one of the most important) is allowing dogs the choice to say yes and no in our interactions with them and respecting that choice. It is common to assume that since dogs are our literal property within the lawful margins of society, we can do what we want to them when we want to do it. Part of this assumption is also linked to being a "responsible owner" and being in effective control (control of their behaviour and their movements). And yet another part of this is the lack of awareness that dogs, too, need to have the right of bodily autonomy. We are not simply entitled to be able to touch dogs because it pleases us. In Chapter 3 I will discuss the use of control and choice and when and how to provide salient options to improve well-being.

1.3.3 Consent and Assent

Consent is an animal (human or nonhuman) giving permission or agreeing to something that involves them. It differs from assent—agreeing that something may happen—by assuming that for consent to occur, there must be the ability to fully understand the stakes. I specially choose the term "consent" over "assent" for a few reasons, but the largest reason is the powerful implications for shifting the exceptionalist narrative and perspectives about how/if consent can apply to nonhuman animals. Science often struggles to apply any human definitions to other animals, claiming it's anthropomorphic, as if that is some type of profane expletive. But as De Waal (2016, p. 26) says, "unjustified linguistic barriers fragment the unity with which nature presents us . . . Our terminology should honour the obvious evolutionary connections." Behaviour studies do not have to necessarily avoid anthropomorphism completely. When applied critically it can potentially enhance the latent reality of our dog's mental lives and acknowledge that dogs do feel, at the very least, similar primary emotions that we experience. Bekoff (2000) says that "anthropomorphism allows other animals' behavior and emotions to be accessible to us" (p. 867).

Conceding that "unthinking" anthropomorphism is not scientific, he argues that "biocentric anthropomorphism" (also referred to as "critical anthropomorphism") is compatible with empirical scientific investigation (p. 867). Especially when considering emotions, Bekoff suggests that using anthropomorphism is mostly harmless, whereas "closing the door on the possibility that many animals have rich emotional lives . . . will lose great opportunities to learn about the lives of animals" (p. 869). In line with Bekoff, James Serpell (2003) also argues that anthropomorphism "is what ultimately enables people to benefit socially, emotionally, and physically from their relationship with animals" (p. 83).

Some may argue that consent given by a nonhuman animal could never be legally binding because we assume nonhuman animals do not have the capacity to decide for themselves. This only undermines the abilities dogs are capable of, whether humans have the capacity to understand them fully or not. Consent doesn't actually require free will or reflexion. It can be defined as a chance to experience assenting to a proposition, or to communicate lack of assent. Any animal that can express themselves can give or withdraw consent. I reiterate here that I use the term **consent** over assent to deliberately highlight that the construction of consent is one entrenched in human bias, limiting our scope to include animal others and instead of bending the established human definition of consent to try to apply it to our interactions with dogs, we need to create a definition that is specific to their experiences. I am not necessarily saying dogs need to learn to speak and write in order to enter a legally binding contract; I am saying that application and definition of consent needs to be modified to fit their own unique set of qualities.

In human terms, consent can be further subdivided into different categories. Dickens and Cook (2015) outline the following:

1. An expression of consent whereby consent is given either verbally, in writing, or through a clear nonverbal gesture.
   a. Potential dog example: An averted gaze and head turn when we reach to touch them is nonconsenting.
2. Implied consent is inferred by one's actions or inactions within a particular context.
   a. Potential dog example: A head nudge and leaning in to garner touching/petting is consenting.
3. Informed consent is permission by someone who has a clear understanding of the facts, implication, and future consequences of an action.
   a. Potential dog example: Having a dog station as a consent signal for nail trims (more information about cooperative care and predictive cues in Chapter 4) is consenting.
4. Unanimous consent is a general consensus given by a group of several parties. This is out of the scope of dogs' social needs.

5. Substituted consent is consent given on behalf of someone else when it's in their best interest.
   a. Potential dog example: I may consent to my dog having surgery because I know it will save her life. This is a decision made on her behalf because there is no way for her to reflect on her options.

Dogs, too, are able to gather information about their environment and the context of situations if we take the time to teach them. However, companion dogs are not fully autonomous, and they are dependent on their human caregivers to make many decisions on their behalf. Dogs also gather information in different ways than humans, but they can use that information to make informed choices in a variety of ways. We can teach our dogs how to use that information to opt in and out of certain interactions, communicate their concerns, negotiate, and ultimately consent in many situations when given the chance. Thus, most categories of human consent are arguably applicable to dogs if we simply adjust our perspective.

That being said, we have created dogs in a way that has made them continually dependent throughout their entire lives, as discussed in length previously. This means, as caregivers, we are also responsible to make some decisions on their behalf (similar to "substituted consent"). But making decisions on behalf of someone else means those decisions should serve that individual rather than our own needs or desires. There will be situations where we may not be able to respect the terms of our dog's decisions, and these situations need to be assessed carefully and thoughtfully to ensure they are in the dog's best interest overall. This might occur when we have additional information that is impossible to convey accurately to our dogs. This makes the concept of consent a complex issue, and one that this book aims to restructure in a way that considers a dog's perspective.

While consent is critical to the understanding of canine well-being as well as human safety (prevention of altercations and defensive behaviours), it remains a nebulous concept. Even for other humans we are often not privy to the details of personal experiences. Knowing the experiences of dogs is far more elusive and difficult to capture or measure, and thus, this area of study is virtually nonexistent or, at best, is fragmented in various texts. In fact, the dearth of literature on nonhuman animal consent echoes the lack of academic attention to this critical model. Even within the literature on nonhuman animal autonomy there lacks consensus on what it us, how it should be defined, or how it is communicated.

1.4   CRITICAL APPROACHES: HURDLES TO DEFINING CONSENT FOR DOGS

There are many hurdles and barriers to applying consent to other animals. We need to consider whether consent can and should apply and in what way? This struggle is largely amplified by the assumptions we make about the reflection and reasoning involved in informed, written, and/or verbal consent and how to apply this concept to nonverbal animal others. But simply not being able to reason or reflect (whether they can or can't and to what degree and in what capacity remains largely unknown) doesn't justify ignoring the consent of others just as it doesn't justify the repudiation of autonomy. But accepting the idea that dogs can consent means there are significant implications for the use of dogs in research and testing (no matter how benign) even to advance our knowledge about their capabilities that may support the argument for consent in the first place. And if dogs are capable of providing valid consent, then it would be unethical to subject them to experimental procedures without their explicit agreement. Moreover, if dogs are used in laboratory experiments, their autonomy would have to be respected, and they should be provided with appropriate conditions to express their natural behaviours and preferences. This is not something humans seem quite prepared for. Despite the advances in the way we think about human abilities, we are still far from a conception of autonomy—and thus of consent—in other animals partly because it isn't always convenient for us. I would argue that if we are to move forward, we need to address the fact that there are absolutely many situations where consent is not at all avant-garde but is necessary if we are to continue to share our lives with dogs.

You might ask, if a dog cannot consent in all situations, how do we evaluate the moral dimensions of what is "best for" our dogs in situations of substituted consent? We often fail to notice other animals' realities, and we are inherently human-biased; yet dogs remain dependent on humans to learn and to be empathetic to their emotional and perceptual needs. Not only do we struggle to see our dogs' realities, but we also struggle to take them seriously—for example, by dressing them in costumes and by breeding them for their aesthetic value despite the cost to their physical health and well-being (for example, French bulldogs who often suffer from severe skin and breathing issues and can rarely have a natural vaginal birth due to the enormity of their head to body ratio). Because humans have a vested interest in using other animals for our own progress and convenience, there is a concern that we may ignore their agency and adopt a more egocentric picture of our dog's requirements and preferences. This is not to say we cannot mutually and compatibly benefit from having dogs in our lives, but we need to recognise the dissonance around consent and autonomy of/for dogs—and not just amongst scholars and philosophers, but within society's discourse about nonhuman animals. Donaldson and Kymlicka (2011) suggest that ones' dependence is not in opposition of independence, but that "recognizing our inevitable (inter)dependence is a precondition to supporting people's ability to express preferences, develop capacities, and make choices" (p. 84). Following this view, we can then improve how we help dogs to do things like choose and express preference in their own way. We fail our dogs not only when we fail to meet their basic needs, but also when we fail to recognise their individuality and the ways in which they can develop a wider scope for greater agency, even micro adjustments, or at least the sense of greater agency in some form.

Another roadblock we face when defining consent for dogs is that consent is a distinctly human-centred concept as we have briefly explored in the previous section. Murray (2017, p. 206) suggests that the morality of consent theory is created by humans, for humans, saying that

> Consent theory sees morality as a compromise among agents voluntarily willing to restrict their own behaviour concerning other persons—only if reciprocated—for the satisfaction of mutually beneficial interests of the bargaining parties. If humans have no similar interests in bargaining with nonhuman animals . . . then the moral sphere that contractarians create does not extend to animals.

Behaviour research indicates that many species of animals do in fact benefit from being with and learning from conspecifics and reciprocate through mutually beneficial interactions with one another. For example, there are studies that indicate how other animals cooperate with one another, and not just other primates, but birds (for example, Heaney et al., 2017; Ortiz et al., 2020), bees (for example, Sharma et al., 2017) and, yes, dogs (Range et al., 2019). Dogs also show empathy toward conspecifics (Quervel-Chaumette et al., 2016) and toward humans (Sanford et al., 2018) and demonstrate human perspective-taking (Catala, 2017) and likely some form or degree of theory of mind[2] (Farhoody, 2018; Horowitz, 2011). We will return to some of these amazing abilities dogs have later in this book to highlight why considering canine consent is more than just a radical notion, but something both tangible and necessary.

1.5 LAYING THE GROUNDWORK FOR A DOG-INDEXED DEFINITION OF CONSENT

Regardless of whether or not other animals experience the world the same way we do, it is not an excuse to do to them whatever suits us. We are seeing a rise in pathologies and dogs who are struggling behaviourally and emotionally in various ways, and a lot of this bears on the pressures and expectations we place on them to suppress their natural behaviours and conform to human rules. Therefore, instead of applying a human concept of consent to dogs and having the terms violated repeatedly, it's important instead to apply a **dog-indexed definition of consent**—one that is specific to their needs and abilities. This definition should account for their species-specific needs, wants, and unique perspective and be one that removes as much human bias as possible.

I hope this book will bring us a little closer. But as you will see, as we take a deeper dive, these issues are much more complex than they first appear. In the last chapter I outline a detailed description of a dog-indexed definition of consent that can be used and applied in all our interactions with dogs—from their physical care and health—to daily interactions such as touching and teaching, learning, and skill building. The overall goal in achieving this will be to include the following steps:

1. Increase (at least the feeling of) autonomous choices. Though companion dogs may not be fully autonomous, the lives of free-roaming dogs can teach us a great deal about the way we manage dogs in our homes and in public spaces and the effect that has on their behaviour. Changes we make for their well-being may include providing the opportunity to make more salient choices and only intervening when it falls under "their best interest" rather than human convenience. It also means the freedom of expression, allowing dogs to fully experience their world and to be able to communicate their needs and wants without fear of reprimand or being ignored.

2. Provide more choices based on understanding individual and species-specific preferences and motivations. Assessing individual preferences and basing available choices off those preferences is a way to help our dogs live fulfilling lives. Situating the environment in ways that favour their ideal choices is preferable when teaching our dogs the social skills they need to navigate their world. Teaching should (and can) be done mindfully and without intentional coercion or punishment. Teaching skills should be aimed at providing our dogs with the information they need to thrive in a human-centric, but shared, society.

3. Teaching skills that help dogs learn the social rules in a way that affords them greater autonomy into their adulthood. Though some may argue that "training" a dog is still a form of control (therefore counterintuitive to being truly autonomous), it's also necessary process for all of us to learn social rules in order to live in a shared environment and negotiate relationships successfully.

4. Learn what consent looks like in dog terms. This means dropping our preconceived ideas about consent and how that might be expressed and respected. Human biases can hold us back in considering our dog's ability to consent using their natural resources, like body language. Information is provided to our dogs through their learning history, so teaching dogs predictable routines is paramount for them to make informed choices.

5. Know when and how to ask for consent and when it might be more important to override our dog's preferences to improve their lives in the future. Obviously, as it stands, humans own their dogs; dogs are *property*. They don't have the freedom to grow into independent adults like some other animals and humans. We continue to manage their lives, and this comes with a certain expectation of personal responsibility. But rather than imposing unrealistic demands on our dogs, we should base any decisions made on their behalf on improving their well-being. When we can offer the foresight of a better overall life, we may have to make

decisions on their behalf, in which case, asking for consent and then ignoring them if they say no will simply deteriorate the trust within that relationship. We will look at how to navigate those difficult situations in the least intrusive way.

If we want to think about ethical relationships with dogs—such as teaching skills to live well rather than training for "obedience" and allowing dogs to consent and live with greater autonomy—then we must do so in the current context of their experiences in a human-centric world. If dogs are considered as an inherently important species in their own right, then there is no "wildness" to shed and no "taming" needed. The binary of wildness and tameness, nature and culture, is obsolete; it obscures the characterisation of dogs and maintains oppressive ideals about other animals (Taylor, 2011). To be fruitful in our endeavours of creating a dog-indexed definition of consent, we have to first remove the cloche of human exceptionalism and rethink how they fit into our society and into our lives.

> **Key Concept:** We need to create a dog-indexed definition of consent that includes choice, autonomy, and consent based on canine-specific communication, canine-specific (and individual) needs and preferences, and the interplay of the canine-human relationship (including dependent and independent decision making).
> **Critical question:** Can dogs have a private life? If they have no private life, can they be considered autonomous?

NOTES

1. The human exceptionalism (HE) paradigm views nature (and all other animals) as less significant and less important than humans. Though HE takes on many forms, the shared assumption is that the human species are the only species that display the mental complexities that include thinking and feeling in any way considered to be important.
2. Theory of mind is the ability to understand and consider someone else's mental state.

REFERENCES

Anderson, K., & Perrin, C. (2018). "Removed from nature": The modern idea of human exceptionality. *Environmental Humanities, 10*(2), 447–472.

Anderson, U. S., Benne, M., Bloomsmith, M. A., & Maple, T. L. (2002). Retreat space and human visitor density moderate undesirable behavior in petting zoo animals. *Journal of Applied Animal Welfare Science, 5*(2), 125–137.

Bekoff, M. (2000). Animal emotions: Exploring passionate natures current interdisciplinary research provides compelling evidence that many animals experience such emotions as joy, fear, love, despair, and grief—we are not alone. *BioScience, 50*(10), 861–870.

BHARCS. (2023, June 9). In recent times, there's been an increase in appreciation of ethological studies on free-living dogs (FLD) [Status update]. *Facebook*. www.facebook.com/bharcs/photos/a.319918418363847/1926856294336710/

Buchanan-Smith, H. M., & Badihi, I. (2012). The psychology of control: Effects of control over supplementary light on welfare of marmosets. *Applied Animal Behaviour Science*, *137*(3), 166–174.

Catala, A., Mang, B., Wallis, L., & Huber, L. (2017). Dogs demonstrate perspective taking based on geometrical gaze following in a Guesser–Knower task. *Animal Cognition*, *20*(4), 581–589. https://doi.org/10.1007/s10071-017-1082-x

Cochrane, A. (2009). Do animals have an interest in liberty? *Political Studies*, *57*(3), 660–679.

Coe, J. B., Young, I., Lambert, K., Dysart, L., Nogueira Borden, L., & Rajić, A. (2014). A scoping review of published research on the relinquishment of companion animals. *Journal of Applied Animal Welfare Science*, *17*(3), 253–273. https://doi.org/10.1080/10888705.2014.899910

Cunningham-Smith, P., & Emery, K. (2020). Dogs and people: Exploring the human-dog connection. *Journal of Ethnobiology*, *40*(4), 409–413.

De Waal, F. (2016). *Are we smart enough to know how smart animals are?* W. W. Norton & Company.

Dickens, B. M., & Cook, R. J. (2015). Types of consent in reproductive health care. *International Journal of Gynecology & Obstetrics*, *128*(2), 181–184.

Donaldson, S., & Kymlicka, W. (2011). *Zoopolis: A political theory of animal rights*. Oxford University Press.

Eddy, T. (2003). What is a pet? *Anthrozoös*, *16*(2), 98–105. https://doi.org/10.2752/089279303786992224

Farhoody, P. (2018). Determinants of the acquisition and display of behaviors associated with "guilt" (appeasement behaviors) by companion dogs. *CUNY Academic Works*. https://academicworks.cuny.edu/gc_etds/2809

Francione, G. L. (2000). *Introduction to animal rights: Your child or the dog?* Temple University Press.

Fudge, E. (2014). *Pets*. Routledge.

Giroux, V. (2016). Animals do have an interest in liberty. *Journal of Animal Ethics*, *6*(1), 20–43.

Gompper, M. E. (2014). The dog-human-wildlife interface: Assessing the scope of the problem. In M. E. Gompper (Ed.), *Free-ranging dogs and wildlife conservation* (pp. 9–54). Oxford University Press.

Gray, J. (2016). *Straw dogs: Thoughts on humans and other animals*. Farrar, Straus and Giroux.

Greenebaum, J. (2009). "I'm not an activist!": Animal rights vs. animal welfare in the purebred dog rescue movement. *Society & Animals*, *17*(4), 289–304. https://doi.org/10.1163/106311109X12474622855066

Gruen, L. (2011). *Ethics and animals: An introduction*. Cambridge Applied Ethics, Cambridge University Press.

Gruen, L. (2021). The moral status of animals. In E. Zalta (Ed.), *The Stanford encyclopedia of philosophy* (summer 2021 edition). http://plato.stanford.edu/archives/sum2021/entriesmoral-animal/

Heaney, M., Gray, R. D., & Taylor, A. H. (2017). Keas perform similarly to chimpanzees and elephants when solving collaborative tasks. *PLoS One*, *12*(2), e0169799. https://doi.org/10.1371/journal.pone.0169799

Horowitz, A. (2011). Theory of mind in dogs? Examining method and concept. *Learning & Behavior, 39*, 314–317.

Hughes, J., & MacDonald, D. W. (2013). A review of the interactions between free-roaming domestic dogs and wildlife. *Biological Conservation, 157*, 341–351.

Irvine, L. (2004). A model of animal selfhood: Expanding interactionist possibilities. *Symbolic Interaction, 27*(1), 3–21. https://doi.org/10.1525/si.2004.27.1.3

Jones, E. E. A. (2022). *Silent conversations: The influence of human exceptionalism, dominance and power on behavioural expectations and canine consent in the dog-human relationship* [A thesis submitted in partial fulfilment of the requirements for the degree of Doctor of Philosophy in Human-Animal Studies, University of Canterbury]. https://libcat.canterbury.ac.nz/Record/3183157

Keralis, S. D. (2012). Feeling animal: Pet-making and mastery in the "slave's friend." *American Periodicals, 22*(2), 121–138. https://doi.org/10.1353/amp.2012.0011

Lord, K., Feinstein, M., Smith, B., & Coppinger, R. (2013). Variation in reproductive traits of members of the genus Canis with special attention to the domestic dog (*Canis familiaris*). *Behavioural Processes, 92*, 131–142.

Manning, B. (2013). *Animal abuse on the rise*. www.nzherald.co.nz/rotorua-daily-post/news/article.cfm?c_id=1503438&objectid=11099735

Marchesini, R., & Celentano, M. (2022). *Critical ethology and post-anthropocentric ethics*. Springer.

Murray, M. (2017). *Morals and consent: Contractarian solutions to ethical woes*. McGill Queen's Press-MQUP.

Ortiz, S. T., Castro, A. C., Balsby, T. J. S., & Larsen, O. N. (2020). Problem-solving in a cooperative task in peach-fronted conures (Eupsittula aurea). *Animal Cognition, 23*(2), 265–275. https://doi.org/10.1007/s10071-019-01331-9

Owen, M. A., Swaisgood, R. R., Czekala, N. M., & Lindburg, D. G. (2005). Enclosure choice and well-being in giant pandas: Is it all about control? *Zoo Biology, 24*(5), 475–481.

Palmer, C. (2010). *Animal ethics in context*. Columbia University Press.

Pangal, S. (2021). *Dog knows: Learning how to learn from dogs*. Harper Collins India.

Perdue, B. M., Evans, T. A., Washburn, D. A., Rumbaugh, D. M., & Beran, M. J. (2014). Do monkeys choose to choose? *Learning & Behavior, 42*, 164–175.

Perlmuter, L. C., & Monty, R. A. (1977). The importance of perceived control: Fact or fantasy? Experiments with both humans and animals indicate that the mere illusion of control significantly improves performance in a variety of situations. *American Scientist*, 759–765.

Peterson, A. L. (2001). *Being human: Ethics, environment, and our place in the world*. University of California Press.

Pierce, J. (2016). *Run, spot, run: The ethics of keeping pets*. The University of Chicago Press.

Quervel-Chaumette, M., Faerber, V., Faragó, T., Marshall-Pescini, S., & Range, F. (2016). Investigating empathy-like responding to conspecifics' distress in pet dogs. *PLoS One, 11*(4), e0152920. https://doi.org/10.1371/journal.pone.0152920

Range, F., Marshall-Pescini, S., Kratz, C., & Virányi, Z. (2019). Wolves lead and dogs follow, but they both cooperate with humans. *Scientific Reports, 9*(1), 1–10. https://doi.org/10.1038/s41598-019-40468-y

Redmalm, D. (2019). To make pets live, and to let them die: The biopolitics of pet keeping. In T. Holmberg, A. Jonsson, & F. Palm (Eds.), *Death matters* (pp. 241–263). Palgrave Macmillan.

Regan, T. (2004). *The case for animal rights*. University of California Press.

Rowlands, M. (2019). *Can animals be persons?* Oxford University Press.

Sanford, E. M., Burt, E. R., & Meyers-Manor, J. E. (2018). Timmy's in the well: Empathy and prosocial helping in dogs. *Learning & Behavior*, *46*(4), 374–386.

Serpell, J. (2003). Anthropomorphism and anthropomorphic selection—beyond the "cute response." *Society & Animals*, *11*(1), 83–100.

Serpell, J. A. (2017). *The domestic dog: Its evolution, behavior and interactions with people.* Cambridge University Press.

Sharma, V., Srinivasan, K., Kumar, R., Chao, H. C., & Hua, K. L. (2017). Efficient cooperative relaying in flying ad hoc networks using fuzzy-bee colony optimization. *The Journal of Supercomputing*, *73*(7), 3229–3259. https://doi.org/10.1007/s11227017-2015-9

Singer, P. (1973). *Animal liberation*. Palgrave Macmillan.

Singer, P. (1986). All animals are equal. *Applied Ethics: Critical Concepts in Philosophy*, *4*, 51–79.

Smith, J. A. (2003). Beyond dominance and affection: Living with rabbits in post-humanist households. *Society & Animals*, *11*(2), 181–197. https://doi.org/10.1163/156853003769233379

Smuts, B. (2001). Encounters with animal minds. *Journal of Consciousness Studies*, *8*(5–6), 293–309. https://psycnet.apa.org/record/2001-07704-015

Stibbe, A. (2001). Language, power and the social construction of animals. *Society & Animals*, *9*(2), 145–161. https://doi.org/10.1163/156853001753639251

Taylor, N. (2011). Anthropomorphism and the animal subject. In R. Boddice (Ed.), *Anthropocentrism: Human, animals, environments* (pp. 265–280). Brill.

Tuan, Y. F. (1984). *Dominance & affection: The making of pets*. Yale University Press.

Whitehouse, J., Micheletta, J., Powell, L. E., et al. (2013). The impact of cognitive testing on the welfare of group housed primates. *PLoS One*, *8*(11), e78308.

CHAPTER 2

A TWO-WAY STREET

UNDERSTANDING COMMUNICATION ACROSS SPECIES BOUNDARIES

2.1 INTRODUCTION

Dogs communicate using a variety of vocal and visual signals that include displacement behaviours, calming or appeasement signals, metacommunication, and stress signals. Dogs' communicative behaviours are highly ritualised—these behavioural cues are used to convey meaning while reducing the need for defensive or aggressive behaviours and are most often used to mitigate conflict. Generally, dogs are phenomenal at reading these signals in other dogs provided they've had adequate opportunity to learn through play and social interactions. This is especially true if they have not had adequate and positive socialisation as a puppy during their critical socialisation window.[1] These types of social interactions can help them learn a subtle conversational interaction that can support de-escalation of stressful situations or conflicts. However, humans often misread canine body language, hampering an effective communication system that would otherwise be effective. Thus, dogs may learn to become overly reactive toward other dogs and people for a variety of reasons throughout their lives, but a large reason is the violation of communication about consent and bodily autonomy by inexperienced dogs and humans.

My dog, Juno, is a great example of how valuable it is to be able to effectively communicate her needs and desires. Juno had a rather dubious first 10 weeks of her life in a rural area with little exposure to anything else. That lack of adequate socialisation during her critical period, paired a very likely genetic predisposition to anxiety and having a "sensitive type" personality, means that she frequently defaults to defensive (aggressive or avoidant) behaviour and often barks or growls in many situations she finds frightening or unpredictable—or at least until she learned better skills to cope. She has learned through acquired skills and through the socialisation process that people and other dogs are safe to be around and that if she communicates, her needs will be respected. The biggest part of this process was teaching her that I will (try to) never let anyone touch her without her permission. This fostered a feeling of safety and predictability. Thus, her suspicion waned, and nervousness dissipated over time and practice, and she is now enthusiastic about being in public places, even busy ones. Even with this newfound level of relief, if an unfamiliar dog (or human) were to approach presenting unsolicited and unwanted interactions, she

DOI: 10.1201/9781003361459-2

most likely will communicate her desire to be left alone. Even from the first sign of their intention, she starts the conversation: Her whole body stiffens, she turns her gaze away, she freezes momentarily, her ears lay flat, and she will lift her front paw, curving her body ever so slightly away from the intruder. Her tail is pinned down between her back legs, wrapping through to her belly, and she often will do a slow tongue flick as if she's wiping a runny nose. These signals combined are very transparent, polite "no" signals. If the approaching dog or person doesn't reciprocate with an indication that they understand and respect her request for space, self-preservation kicks in, and she will sometimes switch to flight or fight (avoidance or escape behaviours or growling or barking in an attempt to create space using threat). Through her vibrant canine language, she has provided a dialogue that she is feeling threatened and unsafe, and it grants them the opportunity to respond.

Are these defensive behaviours "problem" behaviours? Are they something I need to "fix"? First of all, Juno did not solicit engagement, and secondly, she has clearly communicated her need—no social interactions. When her requests have been rejected and she escalates to using an overt and convincing form of communication, it's still acceptable, normal dog behaviour. In fact, the problem isn't dog behaviour at all; the problem is human behaviour or the human side of the conversation. Many dogs communicate with subtle body language that is disregarded or ignored. In situations where a dog has been repeatedly ignored, they will lean to escalate their "voice" in order to convey the urgency of their needs. Consequently, they may start to feel anxious or fearful when people approach in anticipation of the likelihood they will be ignored and forced to interact or participate. Likewise, they may start to feel rage or as though agonistic behaviours are the meliorated or the only remaining option they have to communicate effectively.

When a dog demonstrates aggressive (warning/threat) behaviours such as snapping or biting toward a human, it is almost always the dog who suffers the sometimes terminal consequence. In fact, bites to humans occur often when people have missed or misread less overt communicative signals, often assuming instead that their dog has actually bitten unpredictably. Furthermore, some people believe they are justified to do what they want, when and how they want to their dog at any time and that compliance/submission is what makes a good dog "good" (Jones, 2022). How we frame dogs certainly determines how we treat them but also how we see (or don't see) their language, and largely this can be noted as an effect of the unyielding human exceptionalist paradigm we examined in Chapter 1.

A common idiom used by animal activists is that we need to be the voice for the voiceless. The sentiment is engendered by those who care deeply for animal others and is steeped with good intentions. However, it comes with a heavy grievance: *Other animals have the ability to communicate if we take care to separate ourselves from human exceptionalist thinking and attempt instead to take a dog-centred approach.* Of course dogs don't "speak" in words, which isn't to say they can't learn a large amount of important information conveyed in human languages, but they do communicate using gestures and postures, auditory signals, odours, and touch, like many other species on this planet (Siniscalchi et al., 2018).

Dogs demonstrate a compendium of flexible behaviours when communicating with humans, using many of the same gestures applied during interactions with conspecifics. However, some of these behaviours—such as eye contact, for example—can also have a distinct and dissimilar implication when used to communicate with humans (Topál et al., 2014). So even though we have learned much from observing interspecies interactions, the dog–human relationship is still unique. Adding to the complexity of understanding a whole other species' language, though many gestures are under voluntary control, some are not. For example, if a dog is experiencing anxiety, they release a specific body odour (Bradshaw & Rooney, 2016; Handelman, 2012), one our limited human abilities are unable to detect. Despite being involuntary, it signals to the receiver the inner emotional state of that individual (Handelman, 2012).

Obviously for humans, our primary sense is vision, thus the focus of this chapter is on visual and (some) auditory communicative signals because we are best positioned to learn these cues. As you read through this chapter, bear in mind our inherently feeble ability and human bias to understanding canine communication. Not only is interspecies communication complex, we also have yet to learn a great deal, and we will never know the full capacity of their inner lives, including the nuances of their language.

In the remainder of this chapter, I explore the role of communication and how it is a pivotal component to communicating consent. The roles of human exceptionalism and the subsequent social pressures that prevail often prevent a more canine-centred communication between humans and dogs and thus will be highlighted throughout the following sections. Teaching dogs about consent and applying it is achievable, but we first have to remove the cloak of human bias we intrinsically bring so that we can not only understand what canine body language looks like, but how and why it needs to be considered more earnestly. Barriers to a successful two-way communication system are important to anatomise and disseminate.

Key Concept: Communication breakdown comes from

1. Assumptions or myths about dog–human communication
2. A lack of understanding dog communication
3. A disregard for what they are telling us because it may be inconvenient for us
4. Dogs as possessions and without many rights
5. Not taking dogs seriously when they do not consent or withdraw consent

2.2   COMMUNICATION ACROSS SPECIES BOUNDARIES

In Jean Donaldson's classic book, *The Culture Clash* (2013), she likens dogs living in a human-centric environment to humans suddenly finding themselves in an alien-centric milieu. While it's true, through the process of domestication, selective breeding, and evolution, dogs have become skillful at reading and using a range of ostensive and referential signals with humans and learning many spoken words, they are also outstanding at reading and potentially correctly interpreting our emotional states and certainly using our behaviour (and likely odours) as cues for their own behaviour. However, as Donaldson points out, we are still fundamentally different and understanding that difference is important. Humans are only recently starting to gain a small glimpse into the mental states, emotions, and the unique experiences of dogs. We still have a lot to learn, though admittedly, we have come a long way in the last decade alone. We can use the data provided by cognitive behavioural studies to guide us into making more ethical decisions about dogs and for dogs. But ultimately, I don't need science to tell me I should be kind and compassionate and that doing so builds trust and that trust is built by clear communication and that trust reduces the potential of my dog feeling as though she needs to protect herself from my ignorance.

2.2.1   Canine Body Language

As I have mentioned, dogs communicate through a series of information conveyed through body movements. Learning these movements is about learning an entirely new, unique language that is vastly different than our own. Dogs use their whole body, from nose to tail, and it's nuanced with subtle differences we may not even realise: A wagging tail isn't always an invitation for an interaction; an exposed belly isn't always an invitation for groping hands; a yawn isn't always a response to being tired. Aloff (2018) argues that **one single behaviour or gesture is never enough to precisely interpret a dog's behaviour**. A variety of sophisticated, concise, unambiguous, ritualised social signals communicating threat, submission, and appeasement, as well as emotional underpinnings, have evolved in domestic dogs (Eaton, 2011), known commonly as appeasement signals, displacement signals, and stress signals. It's intricate and complex and really quite amazing when you start to take the time to categorically examine communication in action. I would recommend to anyone to go to the park or the beach and simply observe dogs communicating and engaging with each other and the people in the environment. Watch what they do when people approach. Watch what happens when a child runs over with outstretched arms. Watch what happens when you approach your own dog to pet them as they lie on their bed. Ask yourself, how is this different from when they are the ones approaching you or when you get home from work at the end of the day and they are overjoyed to see you?

Veterinarian and animal behaviourist, Kendal Shepherd (2009), created a tool that is often referred to in the dog behaviour community as the Ladder of Communication (or the Ladder of Aggression) (Figure 2.1). It simply shows the general trajectory of observable behaviours dogs may display when experiencing fear, anger, anxiety,

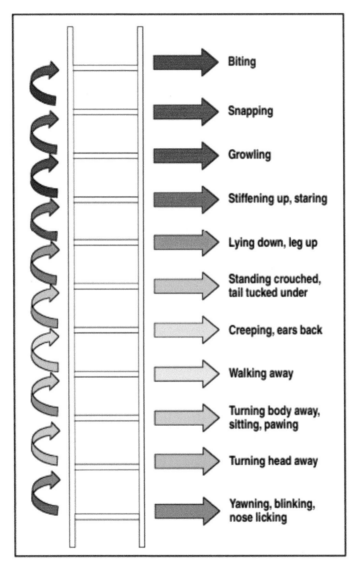

Figure 2.1 The Ladder of Aggression.
Source: Reproduced with permission from the *BSAVA Manual of Canine and Feline Behavioural Medicine*, 2nd edition. The Canine Ladder of Aggression concept was developed by Kendal Shepherd.

pain, uncertainty, or defensiveness and are feeling unsafe or threatened. At the bottom of the ladder, the behaviours we observe are fairly benign but embody a plethora of information and intention. This might be observed as a simple turn of the head, a lowered tail, turning to sniff the ground, a yawn, or a furrowing of their brow—all easily mistaken as rather insipid and inconsequential. As a dog reaches their threshold of tolerance, we might see more obvious behaviours, such as moving away, tail tucked tight through their back legs, a total body freeze, or a recoil. And then, when these signals have failed them, we might see more actionable behaviours, like a growl, snap, or a bite. The aim, most generally, is to avoid harmful social encounters and conflict, not to cause it.

In their most simplistic form, these signals can easily be categorised as "yes" and "no," an applicable binary approach to dog–human interactions. It's not to say some signals are not "maybe" signals or that a dog may not change their mind or need some time to process the conditions of the interaction. But assumption of the former removes ambiguity that can be potentially wrong. In other words, it is better to assume a behaviour is a "no" and respect it or re-ask the question rather than assume it might be a "yes" and risk loss of trust, emotional upset, and potentially dangerous consequences. Yes, it still remains that the largest problem rests in people's ability (or inability) to interpret these signals (Hawkins et al., 2021; Mariti et al., 2012) and learning about what these signals represent is crucial, but putting our own skills into practice is essential for changing how we understand dogs' system of communication.

Let's take a look at the various categories of signals that dogs use and what those mean. Not all signals are a hard no, but all provide us with information about how a dog is feeling, and all should be taken seriously if we are to consider whether an interaction is truly consensual. A dog who is showing signs of uncertainty may need more information, more time, or may need to learn a well-defined system that will allow them to make informed decisions about an interaction before and during an engagement. We will talk more about informed decisions and learned consent signals in Chapter 4.

The following descriptions encompass the fundamental categories of behaviours used to communicate or express emotional states. Be mindful that many behaviours can be seen in more than one category, serving potentially both or similar implicit implications. Likewise, some behaviours may indicate one thing in one situation or another thing in a different context. Some behaviours are two indicators at one time, such as a signal communicating both internal emotional conflict and stress. Therefore, remember these are both complex and somewhat fluid; many communicative systems are ritualised, some are completely involuntary, some are particularly intentional. Therefore, the following is a guideline based on our current understanding of dog communication.

2.2.1.1 Displacement Behaviours

Displacement behaviours occurs when a dog exhibits behaviours that are extraneous to the behavioural context (Breed & Moore, 2016). This is particularly noteworthy because they often occur in socially tense situations. For example, playing with our hair or moving our hands in expressive motions when presenting at a public speaking event might be considered displacement behaviours, even potentially serving some kind of self-soothing purpose. It is theorised to be a strategy used to dissipate energy associated with conflicting motivations, such as the desire to approach a person while at the same time being fearful of that person (or what that person might do). Displacement behaviours in dogs generally manifest as self-grooming behaviours, sniffing, tongue flicking, yawning, scratching, and shaking their whole body as a wet dog might do. These are normal behaviours that a dog may express for various reasons, but to be a displacement behaviour they occur out of their original context. For example, the shaking off that normally occurs when a dog is wet might happen even though they are dry and could transpire directly before or after a stressful interaction occurs.

Many of the signals referred to as "displacement behaviours" are also interpreted as signs of stress, anxiety, or fear, such as tongue flicking (Figure 2.2), self-grooming, and yawning (Figure 2.3) (see Section 2.2.1.4 on stress signals). However, Estep and Hetts (1992) suggest that behaviour may be a communication of fear

Figure 2.2 Juno displaying a tongue flick.

Figure 2.3 Juno displaying a yawn during an interaction.

or stress (referred to by Turid Rugaas in her book *On Talking Terms With Dogs* [2006] as "calming signals") but not necessarily; it may be a matter of who is defining it. Some may consider sniffing as a second dog approaches to be an appeasing signal, a signal used to communicate deference. While others may label this as a typical illustration of a displacement behaviour, where the dog is experiencing conflict (fear and excitement, for example). Now to complicate things; it may be both! Often being in conflict can be somewhat stressful, so this sniffing behaviour may fall into both the category of a displacement behaviour and a signal of stress. Or it could also fall under appeasement signals (which is discussed in the next section). However, it could also be suggested that the dog is simply avoiding conflict with the approaching dog by circumventing social pressure and providing both dogs time to adjust and relax, which may or may not indicate stress and/or appeasement. It is a complicated question, however, and one that is worthy of further inquiry, especially in understanding how communication may be interpreted by humans. A study in 2023 by Pedretti et al. found that displacement behaviours may carry a communicative valence associated with the intention to reduce or avoid aggressive behaviour, but the definitive function of a displacement behaviour was not able to be confirmed. Overall, the dogs classified into the "non-reactive attitude" category showed the same frequency of displacement behaviours when humans approached in both a neutral and threatening way. The researchers noted that it is possible that these "non-reactive" dogs may perceive both threatening and non-threatening approaches by humans to be equally ambiguous, thus using a variety of displacement behaviours in both conditions. This would mean that although it may be likely that these behaviours are used with the intention of communicating their "non-aggressive intentions," their objective use as an appeasement behaviour is uncertain. Aloff (2018) suggests that displacement behaviours may help dogs to manage their personal space within any social interaction and not just during conflict, so this may support the findings of Pedretti and colleagues. However, this doesn't mean that they can't be both appeasement and displacement behaviours (in different or singular contexts), it means we just don't know.

2.2.1.2 Appeasement Signals and Calming Signals

Appeasement signals within the dog behaviour world are commonly defined as a peaceable array of communicative gestures aimed at preventing any antagonistic behaviours that might be anticipated by another individual. Overall (2017) suggests that there are in fact two main definitions of appeasement signals. The first is defined by Rugaas (2006) as specific communicative gestures that are used to calm both parties during potential conflict; thusly she coined the term "calming signal." The second definition is that the use of appeasement signals occur in agonistic encounters and are used to reduce the likelihood of the conflict continuing or escalating (Pastore et al., 2011). Calming signals carry the same communicative value but have a slightly different intention. In *On Talking Terms With Dogs*, Turid Rugaas outlines 30 different (calming) signals. Here is a sampling:

- Turning the head or whole body away
- Yawning
- Making a "soft face;" this includes ears laid back, eyes partially closed, a smooth forehead, and closed mouth
- Freezing, even momentarily
- Lip licking/tongue flicking
- Sniffing the ground
- Sitting or lying down
- Lifting a front paw
- Scratching
- Shake off, where the dog shakes their whole body
- Blinking or squinting (see Figure 2.4)
- Moving slowly

- Slow tail wagging with low tail set or short fast wags rather than broad strokes
- Moving in a curve or with their body curved when approaching or departing
- Lip-smacking

You likely recognise some of these behaviours from Section 2.2.1.1 on displacement behaviours. Whether or not these signals are deliberately communicative signals (for example, a yawn may be instinctual or subconscious when a dog is feeling stressed rather than an intentional form of communication but ultimately communicates conflict avoidance), the important thing is the understanding that the behaviour may indicate the emotional state of the dog. Overall (2017) suggests, however, that evidence of any confirmed appeasement effect is difficult to come by, though starting to pique the interest of some dog researchers (for example, Mariti et al., 2017 discussed next). Overall cautions, however, that the behaviours being evaluated are "not just the signals of 'emotional arousal' but also the physiological processes that contribute to the stress response" (p. v). Furthermore, the behaviours identified most consistently as "appeasement behaviors" are all commonly reported stress-related behaviours as well. So to reiterate my earlier point, there is considerable overlap.

Figure 2.4 Juno displaying blinking/squinting when I ask her if she would like to put her harness on for a walk.

A study by Mariti and colleagues (2017), looked at the use of calming signals (appeasement signals) in dogs in various dog–dog interactions (keep in mind, this may be slightly different when we interpret whether this is similar or the same for dog–human interactions). The researchers recorded 24 dogs while they interacted in pairs with either a familiar dog or an unfamiliar dog. They noted the various calming signals used during these interactions as well as their effects (2,130 examples were recorded in all). They found that the most commonly demonstrated calming signals were head-turning, lip flicking, freezing, and turning away. To demonstrate that these were in fact used as intentional forms of communication, they hypothesised that the dogs would only use these signals when interacting with one another. And that is exactly what they found. The vast use of these calming signals appeared when the dogs were in reasonably close proximity to each other and more calming signals were demonstrated when they were interacting with an unfamiliar dog, rather than a previously known social contact. During this study, they also observed 109 instances of what they classified as "aggressive behaviour." It is noteworthy that calming signals did not precede any of these aggressive behaviours. This suggests that if they had used a calming signal, the likely outcome would have meant that the aggression could be avoided (acting as a cut-off signal). Additionally, in 67% of the instances where aggressive behaviours were expressed, at least one calming signal occurred immediately after, and it appeared to then de-escalated the aggressive behaviour in 79% of those instances.

A study examining two appeasement signals (lip licking and turning away) in dog–human interactions shows that although dogs did use these signals, they actually showed fewer appeasement signals in situations of major threat as opposed to only a minor threat (Firnke et al., 2017). The researchers concluded that this is perhaps because those subtle signs of appeasement are no longer effective as a behavioural buffer in the case of a major threat or a situation that has already escalated beyond the effective use of appeasing signals, and this may be due, in part, to humans' overall lack of awareness of dogs' subtle communication gestures.

Appeasement signals dogs exhibit toward conspecifics appear to be similar toward humans. However, as it stands, this area of study has received little scientific attention, even considering other studies which deal with human–dog interactions. Understanding interspecific signalling is important, especially when we open up the discussion about consent. I hope to see more research in this area in the future.

Critical Corner: If appeasement signals are ignored by people during their interactions with a dog, how does this affect that dog's well-being? Do they feel confident that they will be "heard," or will they advance to more obvious forms of communication (i.e., threats) to de-escalate an unwelcomed assault on their bodily autonomy?

2.2.1.3 Metacommunication

Metacommunication comprises a set of signals used during play as a way to demarcate intentions as "just for fun." These behaviours might include exaggerated bouncy movements, play bowing, and that relaxed-eye, wide-jawed play smile (open mouth, soft face). Because play includes many of the same behaviours demonstrated during conflicts, metasignals are generally thought to be used to relay that the intention is play-related. For example, chasing, biting, growling, mounting, snarling, baring teeth, and the like are all behaviours seen in both playful interactions and in fighting. Along with the assistance of metasignals, dogs will also self-handicap and inhibit their bite force.

Anthropologist Gregory Bateson (1956) called these play-intended signals "metacommunication," which simply means communication about communication. In fact, metacommunication is widely used throughout the Animalia. Humans may smile or laugh when teasing their friend to indicate a playful intent, for example. Similarly, dogs use a play bow to initiate a play session and to communicate their

Figure 2.5 Juno displaying a play bow while playing with me (tug, fetch, chase).

intentions throughout a playful interaction (see Figures 2.5 and 2.6). This may be most useful right before a behaviour that most closely resembles aggression, such as biting or growling. In a 1995 study by Marc Bekoff, he found that dogs will more often use a play bow immediately preceding or directly after a specifically "assertive behaviour" like growling or biting. This suggests that dogs recognise when their behaviour may be misinterpreted as serious aggression.

Metacommunication is a magnificent form of communication that allows for creativity and play through pretending. It has been demonstrated that dogs use behaviours that appear to be one thing but actually mean something entirely separate. Ward et al. (2008) found that play fighting is, for many dogs, one of the primary techniques they use to negotiate and grow new relationships. It's a kind of role-playing used to investigate and learn about one another. Without metacommunication, the outcome could turn into something unintended, but most dogs are aptly equipped to negotiate social encounters (Smuts, 2014).

While metasignals may be directly less important to the topic of consent, they are still important forms of communication. Consent should exist in all interactions we have with our dogs, not just the less desirable ones, like medical care. They can help us to negotiate the "yes" and "no" signals while we play with our dogs. For example, Juno often growls when we play tug together. But her growling is accompanied by metasignals like play bows, a soft face, curling her body, and soft eyes, along with other "yes" signals such as immediately returning the toy to my hands when I let go.

Figure 2.6 Juno's play bow and Percy "shaking off" as a way of diffusing intense play.

It's likely fair to assume that most people wouldn't force a dog to play, though play is often used as a reinforcement, especially in dog sport and detection work. "Yes" signals are an important part of the play interaction, and if play behaviour itself is generally highly valued, we need to be sure that it's not acting as leverage and that it's mutually enjoyable in that particular moment. A dog may choose to engage because it is being offered but not by choice. Metasignals can help to gauge genuine play and thus prevent play turning into conflict or frustration causing stress.

Critical Corner: There is always room to learn. Research by Byosiere et al. (2016b) on the play bow (once thought to be a signalment of play intention) found that both the members of the play dyad are generally still directly before bowing and either chase or play rearing resumed directly after. They suggest that this is used instead as to reinitiate play rather than a communication of intent. Similarly, Byosiere et al. (2016a) investigated the role of play bows in puppies. If play bows signal "just for fun," we would assume to see more offensive behaviours occurring directly before or after. But they, too, found that this was not the case, contrary to Bekoff's 1995 findings. There's still a lot we don't know, and learning is not a linear process.

2.2.1.4 Stress Signals

Signs of stress in dogs may be behavioural operational or may be physical also behavioural, but reflexive though all provide us with information that is relevant. It is worthy to note here that stress, in its biological connotation, refers to a lack of physiological homeostasis and could be good or bad. When you come home from work at the end of the day and your dog is elated with your return, this refers to "eustress" (stress in the form of arousal or excitement). On the other hand, "distress" refers to stress that is perceived as negative (being restrained for vaccinations) (Bienertova-Vasku et al., 2020). For the sake of this discussion, "stress" used throughout this book relates to the latter as a way to avoid further complicating an already intricate topic. All this brings me back to what I, and others, have already said: **One single action or posture is never enough to accurately interpret an animal's behaviour** (Aloff, 2018). A displacement or appeasement activity might indicate eustress, distress, and/or fear—or sometimes not. But all carry important information for us, the observer, and how we may interpret a dog's "yes" and "no" signals when asking for consent.

Stress signals are not necessarily deliberate communicative behaviours but still provide us with information needed for successful two-way communication. This is particularly true in situations that are distressing for our dogs. For example, many dogs are fearful or nervous of being in a veterinary clinic for a plethora of potential reasons. If this is true, the dog may experience more severe distress or prolonged reactions than the stress experienced during a one-off startling noise that occurs in the distance. Though do keep in mind, this is very dependent (as always) on the individual. Also, many of the behaviours listed earlier under displacement behaviours and calming signals may also be stress-related behaviours. Though many of the communicative signals we observe from those two categories may happen because a dog is feeling stressed, stress is actually an internal biological reaction to a threat or an excess demand. When the body's fight, flight, freeze, or fawn system is activated, a surge of hormones rush through the body and activate the parasympathetic nervous system. Because it's a survival mechanism, other systems, like the digestive system, for example, will deactivate so that other areas of the body have the energy needed to deal with the stressful factor. Let's have a look at what these most common physical indications of short-term stress may look like.

1. Anorexia
   An indicator of stress may be a dog who won't eat, even highly valued treats. Of course, the dog may just be distracted or already satiated, but refusing to eat is a good indication of a dog who is feeling stressed given there are no other underlying health conditions.
2. Appeasing Signals
   See Section 2.2.1.2 on these signals.
3. Difficulty Learning
   Dogs who are feeling stressed may seem like they are having a hard time learning something new or focusing on a learning task.
4. Gastro-Intestinal (GI) Upset
   Things like vomiting and diarrhea can be a sign of stress. A good example is a dog who experiences carsickness. This is often a sign that the dog is feeling stressed in the car.
5. Displacement Behaviours
   See Section 2.2.1.1 on displacement behaviours.
6. Connection-Seeking Behaviours
   A stressed dog may seek human contact for reassurance. The dog might jump up, climb up, lean into you, seem "clingy," or crawl into laps.
7. Drooling or Excessive Panting
   As a physical reaction to the gastrointestinal (GI) upset a dog may be experiencing, panting and drooling may indicate stress and accompanying physical discomfort.
8. Excessive Grooming
   This might be licking or chewing paws, legs, their flank, tail, and genitals, sometimes even to the point of self-injury.
9. Hyperactivity and Lack of Focus
   Frantic behaviour, pacing, hypervigilance, pulling, snatching treats, inability to focus are behaviours commonly mistaken as "excited" behaviours. Potentially

these could indicate both eustress and distress, so watch for displacement behaviours that are occurring simultaneously. For example, Juno is often conflicted about visiting the vet. We do a lot of "friendly" visits where nothing unpleasant happens, and she enjoys seeing her favourite doctor and getting treats and attention. She also finds the environment highly stressful. So what often appears to be a "happy" excitement is intertwined with other displacement and calming signals as well as physical stress indicators such as panting and whining.

10. Mouthing
    This may range from gentle nibbling to grabbing, snapping, or biting.
11. Stretching
12. Sweaty Paws
13. Trembling or Cowering
14. Whining or Other Vocalisations (generally high-pitched). While some high-pitched whining may be translated as excitement, a dog who is feeling excited enough to whine is also stressed and potentially anxious, agitated, painful, unwell, or fearful.

2.2.2 Auditory Signals

Dogs have a comprehensive vocal inventory (Yeon, 2007). Humans are able to gather information from dog vocalisations, such as information about their emotional state (Faragó et al., 2010; Pongrácz et al., 2005), and dogs, too, are exceptional at understanding human language. Interestingly, the function of auditory expressions, specifically in dog–human communication, is further evinced by the attenuation of vocal signals observed in free-ranging dogs (Pongrácz et al., 2010), signifying that dog–human social interactions play a very important role to these signals.

The most typical dog vocalisation is the bark, and several studies indicate that barking is an important form of communication for dogs (Feddersen-Petersen, 2000; Pongrácz et al., 2014; Yin & McCowan, 2004). Barking carries information about physical features and emotional states of the signaler, which in turn allows other dogs to discern who the individual is, as well as between the innumerable circumstances they may be experiencing/feeling (Pongrácz et al., 2014). Barking is most often used in close-range interactions such as greeting, threatening, alerting, requesting attention, indicating fear or anxiety, or during play. The acoustic features of barking very much indicate environmental context. For example, some studies show that dogs produce longer barks at a lowered frequency when approached by a stranger and higher-pitched barks more often in isolating circumstances (Pongrácz et al., 2014; Yin et al., 2004).

Growling is another important vocal tool used for communication. Studies show that growls convey meaningful information to other dogs, such as indications of their emotional state, the context of the situation in which the growl is delivered, or physical features of the growling dog, such as size (for example, Faragó et al., 2010; Taylor et al., 2009). Growling is primarily emitted during agonistic exchanges to warn or signal threat outside of the context of play (Yeon, 2007). A good example of this comes from a study by Faragó et al. (2010), who found that dogs tend to inhibit their desire to take a valuable item if they are played a recording of a "guarding growl" first (a growl given in the context of a dog guarding a valuable item, such as a bone). Other important acoustic signals to note are whining (which is an indicator of stressful arousal but also for connection-seeking purposes); howling; and grunts, groans, and squeals (used to indicate acute distress or pain).

A recent study by Albuquerque et al. (2016) examined just how good dogs are at correctly identifying the emotional valence of human vocalisations, as well. It turns out, they are exceptionally good. In the study, dogs were shown human faces and dog faces that were expressing either positive emotions (happy/playful) or negative emotions (angry/aggressive), which were paired with an auditory cue containing either a positive or negative emotional valence. They found that the dogs spent a notably extended time looking at both dog and human faces when it matched the emotional valence of the vocalisation. This indicates that dogs can successfully distinguish between positive and negative emotions expressed by both humans and other dogs. In fact, it has been noted that humans' emotional vocalisations are processed differently in dogs' brains—the right hemisphere is primarily used to process negative emotions (i.e., "fear" and "sadness") and the left hemisphere in the processing of positive emotions ("happiness") (Siniscalchi et al., 2018). To our knowledge, that's likely a cross-species ability that shows us just how incredible the relationship is between people and dogs.

In addition to dogs' incredible abilities for understanding human emotion from hearing our vocalisations and seeing our faces, dogs can process a lot of other information from our voices. For example, they can learn and recognise the symbolism of many human words (Kaminski et al., 2004), and they can use the intonations in our voice as a social referential cue to gather information about peoples' actions and reactions to novel or ambiguous objects and thus change their behaviour accordingly (Merola et al., 2014). While vocal indications may be less important to dogs than visual cues (example, pointing, gaze directing, or orientation) for helping to guide dogs through more vague choice-based circumstances (Gaunet, 2010; Pettersson et al., 2011), the tone of human voices seems to be more effective in communicating our *intentions* (Pettersson et al., 2011).

What about us? Do humans share the same remarkable abilities that dogs demonstrate? Can we also understand the intention, meaning, and emotional valence of dog vocalisations? Existing evidence supports the idea that dogs' vocalisations are indeed effective systems for cross-species communication (Faragó et al., 2017; Pongrácz et al., 2005, 2006), and humans can assess the emotional meaning and context of various growls and barks with a fair amount of accuracy. Humans are also pretty good at cataloguing various barks according to their emotional undertones. For example, one study showed that people rate (human) stranger-directed barking as more "aggressive," while barking during isolation is rated as "despaired," and barking during play interactions as "happier," even when removed from the visual context of the matching behaviour (Pongrácz et al., 2005). This is also demonstrated with the ability for humans to classify a dog's vocalisations even when they have no previous experience with dogs. So while vocalisations clearly carry meaning for dog communication and for our ability to understand the internal mental state of our dogs, often growling for the purpose of threat means that more subtle signals have gone unheeded. So while we never want to punish a dog for growling and we always want to take a growl very seriously, more subtle "no" signals need to also be observed with the same urgency.

To summarise the preceding sections on communicative signals, there are two tables to index examples of how a dog will communicate "yes" and "no" during interactions with humans (Tables 2.1 and Table 2.2). These signals should be considered communication when allowing for consent/consent withdrawal or decline within dog–human interactions. These behaviours may indeed be individualistic and context dependent. But it serves to act as a guide to help people learn canine language and recognise naturally occurring (i.e., not specifically taught) indications of consent. However, these signals can be used with more veracity and consistency in future interactions if the dogs has learned them to be a useful and impactful communicative signal.

Key Points: Displacement behaviours, appeasing signals, and stress signals are fluid. You may find one or more behaviour can fit into different categories depending on the context of the situation, the emotional valence, and the accompanying signals you observe. All communication, including metasignals and vocalisations, are important for consensual communication to occur. Understanding canine language allows us to more accurately interpret a dog's agreement to interact or be handled. Doing so will allow us to avoid overly stressful situations and help to build a successful repertoire of learned consent behaviours (discussed in Chapter 4).

Table 2.1 **Descriptions of "Yes" Signals. It is always important to look at the whole body when assessing and making assumptions about a dog's feelings or emotional state. Most of these behaviours happen in conjunction with other signals. Some behaviours are individual to each dog. These signals may also be used to say "yes" to different activities. For example, a dog who leans into you or paws your arm may want to be pet; a dog who brings you their toy or does a play bow may not want a pet but wants to interact through play.**

YES
Leaning into the person (tactile)
Approaching the person with a relaxed body and soft facial expression

Table 2.1 (Continued)
YES
Fully engaged with the person (not engaged with someone or something else at the time of engagement)
Using their paw to ask for petting (tactile)
Nose or head nudge (tactile)
Some vocalisations directed at the person (auditory), such as a medium to higher-pitched bark that is apparent during play and connection-seeking behaviour
Loose, relaxed body movements/curved body
Tail: loose and relaxed (wagging or not wagging)
Relaxed ear position or forward and engaged
Jumping up on the person (seeking closeness or connection)
Bring a toy to the person for mutual play engagement
Approaching a person when invited into their space
Play bow
Soft, relaxed face (soft eyes and smooth forehead with a relaxed jaw)
Soft eye contact/gazing
Learned "yes" behaviours such as a chin rest or stationing*

Source: *Learned consent behaviours are discussed in detail in Chapter 4.

Table 2.2 **No Signals. Signs of Stress, Uncertainty or Fear. In dog language, these signals mean stop. These are also the signals that humans most often miss or misinterpret. It is advised to watch their entire body as many of these behaviours are seen in conjunction with others in this list. For example, a dog may stiffen, tuck her tail, lift her front paw, flick her tongue, and avert her gaze simultaneously. Others may show a different combination of behaviours. Some of these behaviours, however, could potentially be "maybe" behaviours. Perhaps they are simply needing more time to process the information provided and make a decision. However, "maybe no" should always be considered a no until proven to be a clear "yes." Some dogs have learned that their subtle, conflict avoidance behaviours are repeatedly ignored and feel they must resort to more overt behaviours immediately.[2]**
NO (OR "MAYBE NO")
Sniffing the ground or surrounding areas
Lip flick
Yawn
Lifting and holding up their front paw
Turning head away from person or averting their gaze

(Continued)

Table 2.2 **(Continued)**

NO (OR "MAYBE NO")

Lowered body position or rounded through their back (making themselves appear small and unassuming)

Lowered/tucked tail

Stiff body or freeze (even momentarily)

Trembling

Shaking off

Ears flattened and/or pushed back

Whites of the eye showing ("puppy dog eyes") also known as whale eye

Trying to escape/leave (sometimes this is a very slow and deliberate walk away from the person)

Tightly closed mouth (pursed lips)

Exposing the belly (particularly with one leg lifted and tail wagging with short, quick strokes or tucked)

*Learned "no" or opt out behaviours

THREATS
(OVERTLY "NO")

Lifting the lips to expose teeth

Hard stare

Growling or snarling

Barking or lunging at the person (functioning to create space)

Air snapping in the person's direction

Nipping (contact without breaking skin)

Biting (contact and/or bruising and/or breaking skin). Bites may rage in duration and severity.

Source: *Learned consent behaviours are discussed in detail in Chapter 4.

2.3   THE PRICE OF BEING A DOG IN A HUMAN (LANGUAGE) WORLD

Let's take a moment to loop back to the four reasons for communication breakdown listed at the beginning of this chapter:

1. Our assumptions about interspecies communication
2. Our lack of respect, understanding, or attention to what our dogs are communicating
3. A disregard for what they are telling us because it may not be convenient for us in that moment
4. Dogs as possessions

As I have discussed in Chapter 1, the price of a human exceptionalist paradigm comes at an abundant cost for our dogs. Philosopher and ecofeminist Val Plumwood (2007, para. 3) says that modern human exceptionalism has raw forms, like those discussed earlier (Christian creationism, Cartesian dualism, Enlightenment rationalism), but that it also has numerous subtle forms that "retains important strongholds in contemporary philosophy and science." Plumwood goes on to say that in this subtle domain lies "one of its main symptoms . . . is an obdurate reductionism that resists recognising mind-like qualities in animals, or human–animal continuity in the sphere of mind—which I will call mind discontinuity" which reverberates throughout the years of human exceptionalism in Western culture and society.

There seems to be one central truth: When human lives are viewed as having a "unique value"—sovereign and superior over other animals—it becomes that much easier to slaughter and consume other animals, to confine other animals for traumatic and painful testing procedures or for human entertainment, and, in the case of our companion dogs, to treat as possessions to be owned, designed, loved, discarded, or subjected to whatever suits our needs (or presumably theirs), moulding them to fit into our lifestyles. We paradoxically claim to love our dogs (and surely, we do), though this offers little salve to their objectification and subsequent treatment as "pets" under the rule of their "masters" (despite the change in language, not too much else has really changed). But clearly humans miss out, too, under this design. A genuine connection to dogs can occur when we aim to see the ways they live and experience their environment and the way they may express their thoughts and feelings, even when they may not reflect our own. Not only does this represent the give and take of all relationships, but it also means less conflict and less consequence for both human and dog to conflict.

In an essay by Robin Wall Kimmerer (2022, para. 2–3), she poignantly highlights the dichotomy that human exceptionalism portrays—how it affects the nature we experience and the influence that has on how other animals get to experience nature. She asks,

> What if a single species, out of the millions who inhabit the planet . . . are somehow more deserving of the richness of the Earth than any other? [What if] . . . all the ecological laws that constrain growth and consumption do not apply to us?

She continues by proposing that in a hypothetical unwitting social experiment in worldview (or perhaps less unwitting than we might hope to assume) that examines this supposition, we might ask, "What would happen if we behaved as if the Earth were nothing more than 'stuff'—a strictly materialist, utilitarian view of the Earth—and, moreover, all the stuff belonged to us?"

How does this apply to the discussions of communication? It's a matter of perspective—if we assume that those who speak are more "intelligent" and more deserving than those who don't, do we believe we must be the voice for those who communicate in different ways? Do we assume our "intelligence" places us in a position of power to decide what happens and how and when it happens? And to be frank, I do not think humans are the measure of intelligence; James Brindle writes in his book, *Ways of Being: Beyond Human Intelligence*, that there is a diverse intelligence among all animals (including human animals). However, many tests designed to measure intelligence are based on human constructs of intelligence, and likely many animals of differing capabilities will lack the interest or motivation to even participate in such tasks to begin with, let alone being able to communicate in a way that our limited brains can comprehend. Human exceptionalism and inherent human bias certainly shrouds our ability to learn to really know other animals in a way that is meaningful to them. This is not to say that we should always be permissive to dogs nor assume they always know what's best; it's about a respectful conversation built on trust, one where both participants are heard. As it stands, they exist within a relationship across species boundaries and share our experience in many ways, and in order to live in an ethical relationship with dogs, it's our cardinal duty to learn to communicate with them effectively and with awareness. This, in turn, allows us to share a dog's experience in many ways, too.

Empirical evidence, despite the inherent human biases we bring, has produced some deeper understandings about canine communication, including the myriad of ways they are especially attuned to our cues, emotions, and behaviour patterns as highlighted in previous sections. Some other examples include evidence of dogs demonstrating their proficiency for learning human social cues (example, Oliva et al., 2015), ostensive cues (example, Kaminski et al., 2012; Topál et al., 2014), reading our referential signals and understanding our intentions (example, Bhattacharjee et al., 2018), and our emotions (for example, Müller et al., 2015). We often fail dogs when we don't take the time to verse ourselves in their vast and complex language in the same ways they have for us. It takes practice, great observational skills, and education to fluently learn a very complex system of cross-species communication. And given the sheer amount of dogs who share our homes and neighbourhoods, we really ought to work toward a more multispecies, multilingual society.

NOTES

1. Puppies enter a critical period of socialisation that lasts from ~3 weeks of age to ~16 weeks of age in which critical brain development is occurring, and puppies should be exposed to various stimuli with careful attention to make positive associations.
2. I highly recommend reading the book *On Talking Terms With Dogs* by Turis Rugaas. The book offers excellent photos and descriptions of dog body language.

REFERENCES

Albuquerque, N., Guo, K., Wilkinson, A., Savalli, C., Otta, E., & Mills, D. (2016). Dogs recognize dog and human emotions. *Biology Letters*, *12*(1), 20150883.

Aloff, B. (2018). *Canine body language: A photographic guide*. Dogwise Publishing.

Bateson, G. (1956). The message "this is play." *Group Processes*, *2*, 145–241.

Bekoff, M. (1995). Play signals as punctuation: The structure of social play in canids. *Behaviour*, *132*(5–6), 419–429.

Bhattacharjee, D., Sau, S., & Bhadra, A. (2018). Free-ranging dogs understand human intentions and adjust their behavioral responses accordingly. *Frontiers in Ecology and Evolution*, *6*, 232.

Bienertova-Vasku, J., Lenart, P., & Scheringer, M. (2020). Eustress and distress: Neither good nor bad, but rather the same? *BioEssays*, *42*(7), 1900238.

Bradshaw, J. W., & Rooney, N. (2016). Dog social behavior and communication. In J. Serpell (Ed.), *The domestic dog* (pp. 133–159). Cambridge University Press.

Breed, M. D., & Moore, J. (2016). Chapter 4–homeostasis and time budgets. *Animal Behaviour*, 109–144.

Byosiere, S. E., Espinosa, J., Marshall-Pescini, S., Smuts, B., & Range, F. (2016a). Investigating the function of play bows in dog and wolf puppies (Canis lupus familiaris, Canis lupus occidentalis). *PLoS One*, *11*(12).

Byosiere, S. E., Espinosa, J., & Smuts, B. (2016b). Investigating the function of play bows in adult pet dogs (Canis lupus familiaris). *Behavioural Processes*, *125*, 106–113.

Donaldson, J. (2013). *Culture clash*. Dogwise Publishing.

Eaton, B. (2011). *Dominance in dogs*. Dogwise Publishing.

Estep, D., & Hetts, S. (1992). Interactions, relationships, and bonds: The conceptual basis for scientist-animal relations. In D. Balfour & H. Davis (Ed.), *The inevitable bond, examining scientist animal interaction* (pp. 6–26). Cambridge University Press.

Faragó, T., Pongrácz, P., Range, F., Virányi, Z., & Miklósi, Á. (2010). 'The bone is mine': Affective and referential aspects of dog growls. *Animal Behaviour*, *79*(4), 917–925.

Faragó, T., Takács, N., Miklósi, Á., & Pongrácz, P. (2017). Dog growls express various contextual and affective content for human listeners. *Royal Society Open Science*, *4*(5), 170134.

Feddersen-Petersen, D. U. (2000). Vocalization of European wolves (Canis lupus lupus L.) and various dog breeds (Canis lupus f. fam.). *Archives Animal Breeding*, *43*(4), 387–398.

Firnkes, A., Bartels, A., Bidoli, E., & Erhard, M. (2017). Appeasement signals used by dogs during dog–human communication. *Journal of Veterinary Behavior*, *19*, 35–44.

Gaunet, F. (2010). How do guide dogs and pet dogs (*Canis familiaris*) ask their owners for their toy and for playing? *Animal Cognition*, *13*, 311–323.

Handelman, B. (2012). *Canine behavior: A photo illustrated handbook*. Dogwise Publishing.

Hawkins, R. D., Hatin, B., & Révész, E. (2021, December 7). The accuracy of canine vs human emotion identification: Impact of experience through dog ownership and belief in animal mind. CABI https://doi.org/10.31234/osf.io/hs8w3

Jones, E. E. A. (2022). *Silent conversations: The influence of human exceptionalism, dominance and power on behavioural expectations and canine consent in the dog-human relationship* [A thesis submitted in partial fulfilment of the requirements for the degree of Doctor of Philosophy in Human-Animal Studies, University of Canterbury]. https://libcat.canterbury.ac.nz/Record/3183157

Kaminski, J., Call, J., & Fischer, J. (2004). Word learning in a domestic dog: evidence for "fast mapping". *Science*, *304*(5677), 1682–1683.

Kaminski, J., Schulz, L., & Tomasello, M. (2012). How dogs know when communication is intended for them. *Developmental Science*, *15*(2), 222–232.

Mariti, C., Falaschi, C., Zilocchi, M., Fatjó, J., Sighieri, C., Ogi, A., & Gazzano, A. (2017). Analysis of the intraspecific visual communication in the domestic dog (*Canis familiaris*): A pilot study on the case of calming signals. *Journal of Veterinary Behavior*, *18*, 49–55.

Mariti, C., Gazzano, A., Moore, J. L., Baragli, P., Chelli, L., & Sighieri, C. (2012). Perception of dogs' stress by their owners. *Journal of Veterinary Behavior: Clinical Applications and Research*, *7*(4), 213–219.

Merola, I., Prato-Previde, E., Lazzaroni, M., & Marshall-Pescini, S. (2014). Dogs' comprehension of referential emotional expressions: Familiar people and familiar emotions are easier. *Animal Cognition*, *17*, 373–385.

Müller, C. A., Schmitt, K., Barber, A. L., & Huber, L. (2015). Dogs can discriminate emotional expressions of human faces. *Current Biology*, *25*(5), 601–605.

Oliva, J. L., Rault, J. L., Appleton, B., & Lill, A. (2015). Oxytocin enhances the appropriate use of human social cues by the domestic dog (Canis familiaris) in an object choice task. *Animal Cognition*, *18*, 767–775.

Overall, K. L. (2017). Appeasement, calming signals, and information capture: How do our subjects tell us what matters to them? *Journal of Veterinary Behavior: Clinical Applications and Research*, *100*(19), v–viii.

Pastore, C., Pirrone, F., Balzarotti, F., Faustini, M., Pierantoni, L., & Albertini, M. (2011). Evaluation of physiological and behavioral stress-dependent parameters in agility dogs. *Journal of Veterinary Behavior*, *6*(3), 188–194.

Pedretti, G., Canori, C., Biffi, E., Marshall-Pescini, S., & Valsecchi, P. (2023). Appeasement function of displacement behaviours? Dogs' behavioural displays exhibited towards threatening and neutral humans. *Animal Cognition*, 1–10.

Pettersson, H., Kaminski, J., Herrmann, E., & Tomasello, M. (2011). Understanding of human communicative motives in domestic dogs. *Applied Animal Behaviour Science*, *133*(3–4), 235–245.

Plumwood, V. (2007). Human exceptionalism and the limitations of animals: A review of Raimond Gaita's the philosopher's dog. *Australian Humanities Review*, *42*, 1–7.

Pongrácz, P., Molnár, C., & Miklósi, Á. (2006). Acoustic parameters of dog barks carry emotional information for humans. *Applied Animal Behaviour Science*, *100*(3–4), 228–240.

Pongrácz, P., Molnár, C., & Miklósi, Á. (2010). Barking in family dogs: An ethological approach. *The Veterinary Journal*, *183*(2), 141–147.

Pongrácz, P., Molnár, C., Miklósi, Á., & Csányi, V. (2005). Human listeners are able to classify dog (*Canis familiaris*) barks recorded in different situations. *Journal of Comparative Psychology*, *119*(2), 136.

Pongrácz, P., Szabó, É., Kis, A., Péter, A., & Miklósi, Á. (2014). More than noise?—Field investigations of intraspecific acoustic communication in dogs (*Canis familiaris*). *Applied Animal Behaviour Science*, *159*, 62–68.

Rugaas, T. (2006). *On talking terms with dogs: Calming signals* (2nd ed.). Dogwise Publishing.

Shepherd, K. (2009). Ladder of aggression. In D. Horwitz & D. Mills (Eds.), *BSAVA manual of canine and feline behavioural medicine* (pp. 13–16). Wiley.

Siniscalchi, M., d'Ingeo, S., Minunno, M., & Quaranta, A. (2018). Communication in dogs. *Animals*, *8*(8), 131.

Smuts, B. (2014). Social behaviour among companion dogs with an emphasis on play. In *The social dog* (pp. 105–130). Academic Press.

Taylor, A. M., Reby, D., & McComb, K. (2009). Context-related variation in the vocal growling behaviour of the domestic dog (*Canis familiaris*). *Ethology, 115*(10), 905–915.

Topál, J., Kis, A., & Oláh, K. (2014). Dogs' sensitivity to human ostensive cues: A unique adaptation? In *The social dog* (pp. 319–346). Academic Press.

Wall Kimmerer, R. (2022). *How the myth of human exceptionalism cut us off from nature.* https://lithub.com/robin-wall-kimmerer-humans-nature/

Ward, C., Bauer, E. B., & Smuts, B. B. (2008). Partner preferences and asymmetries in social play among domestic dog, *Canis lupus familiaris*, littermates. *Animal Behaviour, 76*(4), 1187–1199.

Yeon, S. C. (2007). The vocal communication of canines. *Journal of Veterinary Behavior, 2*(4), 141–144.

Yin, S., & McCowan, B. (2004). Barking in domestic dogs: Context specificity and individual identification. *Animal Behaviour, 68*(2), 343–355.

CHAPTER 3

THE ROLE OF CHOICE AND AGENCY

3.1 INTRODUCTION

Juno walks along the path, smelling each singular blade of grass, one nostril and then the other moving in opposing directions, tracking the origins of what information they hold—in that moment, her world exists in the system of a varying odours and sounds. A series of loud buzzing is omitted from the overhead wires that I cannot hear but assaults her senses in sometimes unpleasant ways that I cannot experience. Her environment must appear (smell, look, and sound) tremendously distinct from my own—the intensity, the movement of time, perception of space, and the regard of the things of sensory worth. Her experiences of this world are what von Uexkull (1957) terms as "*Umwelt.*" That is not to say that there is not a shared experience of the world between us; certainly more so than a fly and a shark, but not as much as another human. That doesn't mean we can't be empathetic to one another's lives; it simply means we have to pay attention and adjust in a way that doesn't always come organically to us.

Concepts of animal liberation have been explored for decades and support several exceptional theories of animal rights. Much of the focus, however, has historically been on eliminating suffering in animal others, which is of course a worthy cause. However, there has been a noticeable gap between the effort to end the suffering caused by slaughterhouses and medical testing labs and considering the emancipation of those we call "pets." To be clear, I am not talking about an abolitionist perspective, necessarily. As it stands, our companion dogs have been selectively bred and conditioned in many ways to be dependent on humans, and an abolitionist perspective is, in our current state of coexistence, irresponsible at best. However, increasing attention to their agency, focusing on the experience of fully functioning adult dogs with fully learned skills, is an important trajectory worthy of a deeper investigation. Additionally, literature on the topic of animal liberation has also historically been less focused on the subject of agency or certainly the topic of consent, so this is somewhat new terrain to navigate.

In Martha Nussbaum's new book, *Justice for Animals: Our Collective Responsibility* (2023), she writes about something called the "Capabilities Approach." First, this approach suggests that, from a normative perspective, the systematic treatment of other animals is morally untenable. Secondly, the Capabilities Approach centres on

DOI: 10.1201/9781003361459-3

the recognition and value of nonhuman animal agency, which is frankly a refreshing perspective (despite some of her more radical claims in other areas of moral episte-mology, perhaps using a reductio ad absurdum aimed at attacking moral realism). Nonetheless, she envisions a "multispecies world in which all have opportunities for flourishing" (p. 191). I strongly relate my argument for defining canine consent to the work of Donaldson's and Kymlicka's (2011) in their book, *Zoopolis*, where they suggest a political theory and practice of animal rights that focus on the intrinsic value and interests of nonhuman animals as some arrangement of citizenship (or "denizens") of a **shared society**. If we shift the current paradigm of dogs as ours to manage, shape, and control, we may be able to shift how we interact with them in a way that benefits their well-being, not simply our own.

The unique experiences of each individual dog, their *Umwelt*, can influence their emotional well-being. Who are we to suggest we know better for them? Perhaps there are times I can use my gift of foresight to prevent Juno's future suffering, but there are times when that same strategy may hinder her ability to behave in a way that best serves her. For example, if I avoid Juno engaging in dog–dog social interac-tions without letting her converse with the other dog properly, I may avoid potential conflict. However, this prohibits her from learning the skills she needs to navigate these interactions with confidence in the future or to have a welcomed social inter-action when she desires it. I am not saying environmental management isn't helpful for a dog who struggles emotionally or who hasn't fully learned the skills needed to navigate certain situations, but I am suggesting that micromanagement may be the root of some of the increasing "behavioural issues" our dogs are experiencing as we attempt to shape them into the human ideal, devoid of agency, choice, and consent.

3.2   THE SCIENCE OF CHOICES

Choice is implicitly and arguably one of the most important components of consent. Having the choice to participate or the choice to withdraw that previous agreement at any point is crucial to feeling safe. The word choice has multiple implications. Choice may refer to the array of objects, things, or situations from which to choose. It may refer to the act of choosing or the process that occurs when presented with options based on personal preference. Or it could refer to that which is chosen. Yet another meaning of choice is "the power or liberty to choose" (Martin et al., 2006). So what is it that we are talking about here? Well, all of it is relevant and leads us to the ultimate question of whether dogs have real choice and when?

Part of answering this will be to examine what sort of stimulus arrangements con-stitute choices. Not all choices are created equal. Choice involves multiple stimuli and multiple discriminative stimuli, but they may not all be equally motivating or even salient.[1] If we think about it, "an organism can choose among alternatives only to the extent that stimuli are correlated with their availability" (Catania & Sagvolden,

1980, p. 77)—thus, the availability of choices offered and their value. Catania (1975) suggests there are distinctions between "free choice" and "forced choice" situations. Identifying the differences may provide a way of recognising the conditions that only "technically" provide a choice even though they do not carry equal or tenable results, such as a person offering a choice of a treat for holding still or a zap from a shock collar and distinguish those from salient choices that offer two or more desired options.

Therefore, in order to provide fair choices, an assessment of preferences is needed. Martin et al. (2006) says that "preference is the relative strength of discriminated operants. Researchers often measure preference as a pattern of choosing" (p. 4). Therefore, presenting salient choices based on previously evaluated preferences is important in the context of supporting dogs who lack the ability to "voice" their choices in a human-comparable way, bearing in mind that preferences are not static and may change depending on motivating operations, options, and distant antecedents, for example, which we will dive deeper into in the following sections. Choices, which are frequently viewed as a vital characteristic of a quality life (e.g., Wehmeyer & Schwartz, 1998), are unreasonably scarce for dogs who can't verbally ask for things and who may not be able to obtain things on their own and are disempowered by human control.

3.2.1 Motivation

In the process of making decisions, choices are based on both the availability of salient options but also individual motivation. Importantly, when we consider how and when dogs make certain choices, we need to look at some of the cognitive and behavioural research investigating motivation, or the desire to engage in a behaviour. Motivation itself can be positive or negative, and it can fluctuate in valence. Motivation is positive for tasks and consequences that a dog wants to engage in and negative for undesirable tasks and outcomes. In many cases, inferences about motivation can be made from observing a behaviour, whether it is approach-behaviour (positive valence) or avoidance-behaviour (negative valence) (Kirkden & Pajor, 2006). Some of these differences in motivation can explained by the shifting individual or environmental influences such as life stage, learning history, or resource predictability. Motivation may also be influenced by an individual's experience of distant antecedents (medical, physical, or nutritional variables). For example, a dog who is tired may feel less motivated to go for a walk. In Juno's case, if it's raining, she is almost always going to be less motivated to go for a walk, though I still always ask. Or a dog who is in pain may really want to walk upstairs to go to bed but is less motivated because it hurts. Additionally, some elements in the variations in motivation can be ascribed to intrinsically fixed traits, such as canine-specific characteristics (genetic or physiological) such as predatory behaviour. Well, all know at least one dog who is highly motivated to chase a rabbit and with a lowered motivation to come back when called during such events.

3.2.2 Preferences

Preferences are predicated on an ability to choose from a contemporaneous set or sequence of available and accessible choices that satisfy the same motivation and offer the most appealing consequence (Amdam & Hovland, 2011). Preference may be individual (for example, I prefer coffee to tea) and will depend on the strength of that motivation over other options (if I have to go to the store to get the coffee because we are out of coffee beans, I may choose a tea instead). In judging these shifting preferences for Juno, I must make inferences about her preferences in the moment they occur based primarily on her observable behaviour. Though past preferences might influence my conception of what might motivate her, I must also gauge her preferences based on current motivation as well. For example, if I pet her and then pause, does she paw my hand to ask me for more or does she lay down and go to sleep? I may initially pet her because she *generally* has displayed a preference for touching in similar conditions, but I ask her what her preference is in this particular moment because preferences fluctuate based on motivation. Another example is Juno's favourite walking spots. She often takes me to the beach in the mornings. She appears to enjoy social interactions with other dogs, scavenging for crumbs kids have left, and sniffing the drift wood that has washed up on the high tideline. The other day, as we approached the beach with an eager stride (positive valence), it began to rain quite heavily. Juno immediately turned around towards the shops (negative valence) and we instead walked under the shelter of the building overhangs. Though her preference on most days is the beach, there were new motivating operations that influenced her preference in that moment (staying dry and sheltered). I am happy that she understood she had a choice and that I would listen to her decision.

> **Key Points:** When we consider the options we provide for our dogs, or describe learning systems that purportedly provide choices, we should reflect on the following:
>
> 1. What are the accessible selections we are offering? Are there numerous opportunities for positive reinforcement, or are the available options only between earning a reward and getting nothing (or getting something undesirable)?
> 2. How are we controlling and regulating the choices they can make?
> 3. Are there choices available that provide dogs various avenues to express their preferences and have some agency in their lives?

3.3   DO DOGS HAVE TRUE AGENCY AND FAIR CHOICES?

Providing our dogs with increased options for choice—choices based on their individual preferences and current motivation—is one way we can create some semblance of agency and increase their well-being. However, as we have discussed, companion dogs live in some degree of perpetual paternalistic relationships with humans and thus are not truly autonomous, though may be provided some agency or conditional agency in some situations. However, dogs, on their own merits and as individuals with their own distinct qualities and abilities, can live autonomously (recall the discussion about the autonomy of street dogs). As I discuss in Chapter 1, I may aim to increase a *sense* of agency by giving Juno more time off the leash or increasing her available salient choice of toys or activities. However, even if I resolve to allow her the most amount of freedom I can, it's still *conditional* freedom, and she makes choices based on those conditions with her *conditional* agency. It's conditional on my power to grant her those freedoms when and where I see fit. Arguably, companion dogs may not be true agents of their own lives in the absolute sense, but we can facilitate better options and allow at the very least established choices about consent in everyday interactions with humans in order to improve their mental wellness.

Mental wellness is contingent on having a sense of autonomy and control over our conditions and our body. This is not just true for humans but for other animals, including dogs. What does the science tell us about the importance of having a sense of control? Veterinarian Frank McMillan (2019) discusses that a vast body of research shows that having a sense of control over one's own life conditions and experience, and specifically over "events or stimuli that are unpleasant," is strongly correlated with feelings of optimism and mental well-being. Additionally, Lagisz et al. (2020) suggest that having and making choices can improve morale and optimism. In particular, having salient choices may be significantly valuable when a dog is faced with an aversive stimulus. Think about a dog who is fearful of other dogs (thus making other dogs an *aversive stimulus*[2]). If that dog perceives they have some control over proximity, intensity, and duration of an aversive interaction, the experience is more manageable and less aversive overall. Having the ability to exercise control over our environment can be rewarding in and of itself and can even increase certain behaviours in the future. As Susan Friedman (2020) says, having a choice is inherently reinforcing and should always be the aim of any learning interaction between humans and dogs. In fact, McMillan (2019) cites several studies of captive rodents to highlight this point, saying that they "exercise control virtually any chance they get" (p. 142) and observes that they find it to be inherently gratifying when they have an elevated control over their environment. I know I feel pretty happy if I can choose to send an unknown caller to voicemail or if I can opt to avoid a busy store and order online instead.

However, if humans orchestrate the lives of dogs in such a way that leave them with limited control over their environmental liberties, social interactions, and the fundamental components of everyday existence, where does that leave dogs? It is us humans who choose where dogs will live, with no independence to leave if they are unhappy. Beyond not being free to choose their friends, home, or human family, they also don't have much choice about eating, mating, how to spend their time, or raising their offspring, for example. These are enormous issues that are entangled within the human exceptionalist paradigm, but their loss of agency occurs in smaller infractions in everyday situations as well, such as the ability to choose their friends, when to use the toilet, or even how far or fast to walk. This lack of agency is exemplified in the burden of a harness or collar and lead on the ability to move freely or the uninvited touching by strangers and the inhibition of inherent behaviours through training programmes (regardless of the technique). Positive reinforcement training doesn't escape without critique, though it may be the kinder and more humane option over abusive tools and techniques that employ positive punishment, for example; though it still has the potential to be coercive if salient choices are not presented with consideration to motivation and preferences. This is not to say we shouldn't teach our dogs the skills they need, but the aim should be choice and agency, and the skills should reflect their needs to thrive in a human-centric environment.

However, having some control over dogs in our care is not necessarily malevolent. It is part of the reality from the selective breeding of dogs for their dependent traits, and if we are committed to being responsible for their well-being, this needs to be considered. Some forms of control can be a lifesaving, humane choice. And as Milligan (2017) suggests, it can be appropriate to use our human gift of forethought to help dogs in certain situations that may benefit them in the future. That is, after all, an inherently important component to any healthy relationship, helping others to succeed and to be happy and healthy. How we should make these decisions on behalf of our dogs is discussed in further detail in Chapter 5.

The subject of agency is one that could be in itself an entire book and perhaps beyond the scope of this one. The complexities of autonomy and power dynamics make it difficult to tease apart how the future of our relationships with dogs might look. But one thing I do want to point out is that while the discussion on agency is an important one to broach, it may not even matter to the subject of consent in the way we may think. **Consent should matter whether or not a nonhuman animal has agency; whether or not a nonhuman animal can be autonomous.** As I wrote in Chapter 1, consent should be the right of everyone. As we are now aware of the parallels between human oppression and other animal injustices, consent will hopefully become normatively imbued as it begins to become rooted in our everyday discourse. Let's shift the focus to the role that consent plays in moral interactions by first examining the autonomous lives of street dogs.

3.4   FREE-RANGING DOGS AND AUTONOMY

"Street dogs," or free-roaming/ranging or free-living dogs, have something that many of our companion dogs do not: Freedom and independence. They can choose when to eat and sleep, and they socialise when and how they choose. They can decide their location and their daily schedule, move about without constraint and limitations, and react in ways that best serve their needs. Therefore, one area of research that could lead to the quintessential concept of who is "dog" is through the study of free-roaming, stray, and/or feral dogs. This can help us to understand the ways that having choice and agency and asking for consent in dog–human interactions is important to the well-being of dogs in our care.

Obviously, every dog—whether living under human control or independent of most direct human influences—will have a variety of subjective experiences within their lives that influence their behaviour, choices, and emotional well-being. However, free-roaming dogs may provide a unique set of behavioural data as they exist across a broad spectrum of independence from humans and may help us to understand dogs living unrestricted by direct human management. Studies on free-ranging dogs can tell us how they generally choose to spend their time and what, if any, behavioural differences may be present. For example, a recent study by Griss et al. (2021) showed that high activity and a greater time spent resting is observed more frequently in companion dogs than in free-roaming dogs, while their moderate activity level was lower. In a study by Corrieri et al. (2018), free-living Bali dogs (those not associated with any one human home) were found to be less active, excitable, and aggressive toward other dogs/other animals than local companion dogs living in human homes. Free-living Bali dogs were also less inclined to chase other animals, dogs, or humans compared to free-roaming Bali dogs who are also living as human companions in homes. Such studies and future studies on free-roaming dogs may potentially give us a deeper knowledge of the variations in the subjective experiences of different groups of dogs. One such project is the Free-Ranging Dog Project whose research focuses on the effects of domestication on dogs' social behaviour, personality, physical, and social cognition (see www.vetmeduni.ac.at). In addition, these studies could help us to investigate the influence that humans may have over the experiences that shape the way dogs both feel and behave.

Congruent to the findings listed earlier, Pierce and Bekoff (2021) argue that the behaviour of free-roaming and feral dogs might be able to forecast how dogs may live and behave in a post-human world—an allegorical interpretation of a world without humans. They discuss how dogs may adapt to a variety of environments both in form and function. It might be a difficult scenario for us to imagine. Speaking personally, my life would never feel complete without dogs in my home. But I believe it prudent to ask, does the same go for my dog when considering me? Maybe for our existing purpose-bred companion dogs, a life with humans is necessary and/or wanted, but what about the future generations of dogs in a world devoid of humans? How might they adapt to living a life without human interference? I think that imagining a post-human world where dogs live as free agents, masters of their own lives, and living with free will, is one way to address the vexing questions about canine agency and defining "who is dog". Defining "who is dog" (apart from humans) can allow us to shape a definition of consent that is canine-specific and relevant to their experience, their "Umwelt."

Pierce and Bekoff (2021) go on to say it is likely that many companion dogs who have been selectively bred and raised for specific human-desired traits and behaviours would not fare well in a post-human world until there was some evolutionary reformation to create an evolved type of dog who thrives in their independence. As it stands, perhaps dogs rely on our care and concern to live a good life in a human-centric, though shared, society. It is difficult to say how many of their abilities (learning, problem-solving skills, etc.) are directly and indirectly affected by humans and how these affect their subjective truths, though it's quite certain we do have an influence on how those abilities develop. And we do know that the ecological niche that companion dogs currently inhabit (i.e., human homes) most certainly influences the evolution and adaptation of their cognition in some ways (Byrne, 1995).

It would be imprudent to ignore the exceptional research being done on free-roaming dog behaviour—behaviour that is uninhibited by the direct influence of human caregiver interactions. The liberty and autonomy that free-roaming dogs experience comes with abundant and rich mental stimulation, problem-solving capabilities, and the free use of species-appropriate communication—all behaviours they learn from conspecifics, without human manipulation contrived of antecedent arrangements and consequences, behaviour modification programmes, and superfluous management. For example, many companion dogs (not all, of course) spend a significant amount of time alone and/or in confinement (kennel, house, fences, leashes), with only a few hours of time to walk/explore, socialise, play, and eat, thus affecting how they choose to spend their "free time." And whether or not we allow them independence, they are still only under the guise of having any liberty. Their designed dependency on humans means that our lifestyle, knowledge, and actions impact their ability to make choices (or not), have agency (or not), consent (or not), learn social skills (or not), and behave naturally (or not). Dogs who spend their days in confined spaces or on the end of (particularly short) leashes and those waiting for a few moments of social engagements or to be granted the opportunity to use the bathroom obviously have much less autonomy (arguably, none), but they also have little space or option to consent in daily interactions with people.

Dogs who live freely certainly have some freedoms and lifestyles that provide them with "a good life" in many ways, but that is not to say that their lives are always fundamentally better. In an earlier article I wrote that it is also worth noting that,

> free-roaming dogs do face challenges, of course, such as living potentially shorter lives because of disease, parasites, and potential environmental hazards as well as the developing and changing ecology (Paul et al., 2016). They may also be liable to experience mistreatment or may go hungry for an unfavorable amount of time. Even humans have

been shown to negatively impact lifespan of street dogs by way of vehicular fatalities and murder/brutality (Paul et al., 2016). These issues could very well impact both activity and behavior and, undoubtedly, more research is needed. Regardless, we can learn a lot from the types of behaviors observed in dogs who are largely uninhibited by humans.

(Jones, 2021, para 8)

Connecting this discussion back to our topic of canine consent, the utilisation of a multidisciplinary approach to investigate the subjective lives of dogs can contribute evidence that both supports consent-based teaching-learning paradigms and that can be used to aid in a thriving coexistence with less compromise and more compassion. As caregivers, providers, and teachers, we come to understand our dog's personal preferences, which can help us to make educated choices for our dogs including when to allow them to make more of their own choices. That is where research on the behavioural ecology of free-living dogs may better our perspectives. Such an improved understanding of canine subjectivity can also allow us to ask valuable questions and find gainful answers that contribute to how we teach and communicate with the dogs in our lives. Though our companion dogs and free-ranging dogs live in separate environmental and social spheres, examining dogs that are uninhibited by direct human influence and control may uncover some key resemblances or contrasts in their behaviour, skillset, and wellness. As research in this area grows and more anthrozoological and behavioural data is collected, we may find that we have a lot to learn and a lot to gain to help our companion dogs thrive.

3.5 MAKING SPACE FOR CHOICE, AGENCY, AND CONSENT

How can we make space in our dogs' lives for them to have greater agency? Empowering dogs is the foundation of helping them thrive and that comes from reclaiming some power and control over their own lives and actions. In fact, it's logical to assume that, given the chance, dogs might make different and sometimes more competent and relevant choices for themselves in contrast to the ones we make on their behalf. By making choices for themselves, dogs have opportunities to acquire new knowledge, develop and hone their skills, learn ways to communicate successfully, and build trust with the people and other animals in their lives. These choices can be based off of their unique innate abilities and also skills that have been learned through various processes of operant and associative learning processes. These considerations need to be at the centre of how we nurture empowerment, confidence, and ultimately, increased agency through consent-based interactions.

Creating the space for agency, choice, and consent can be challenging for many people to conceptualise because it hasn't traditionally been a part of the discourse concerning dogs. Evidence suggests that other animals can, and do, make choices for themselves, measured through nonverbal referencing (Kirkden & Pajor, 2006), but it is debatable how many of these choices are influenced by ostensive human behavioural cues (Marshall-Pescini et al., 2012; Prato-Previde et al., 2008; Range et al., 2009). A better understanding of dog-specific behaviour with the intention of improving their well-being could benefit how dogs are integrated more compassionately into our society in the future, and part of this might be to teach dogs to follow environmental cues and to teach them skills that can be used conceptually once fully attained rather than focusing on dogs who simply follow human cues and direction (which is also an important skill).

Yeates (2018) argues, and I agree, that deliberately managing situations that empower animals to make better choices should be prioritised, which sounds a bit ironic but is what needs to be focused on for dogs in a human-controlled learning/living environment. Examples of this might include teaching consent behaviour for cooperative care techniques or even setting up the environment to make a favourable choice the more desirable option. By making one behaviour more desirable than another, we are simply designing an environment in which the desired behaviour becomes relevant, effective, and efficient (and the undesired response becomes ineffective, irrelevant, and inefficient). For example, Juno might initially want to bark at children on scooters, but I know it's in her best interest to learn to instead simply move away and observe from a safe distance. I can replace the behaviour of barking with a new behaviour of turning off the footpath and make this the desirable option by rewarding her with something she desires (like food). Eventually, she learns the skill of moving over onto the grass when she sees a child on a scooter and this provides her a feeling of predictability and safety. It is the more effective, efficient, and relevant behaviour in that situation. This process is known as differential reinforcement of an alternative behaviour (in other words, I have reinforced the behaviour that is the more desirable option). Because of this, she also now has important information about what is happening and what she should do, and keeps her safe and out of the way. Dogs can make better choices when they have adequate information and time to process information, when conditions are favourable, when they are not enduring undue stress, and when they have learned adequate skills through humane, trust-building methods that follow the Humane Hierarchy of Behaviour Change (which we will take a look at in Chapter 4; see Figure 4.2 for a breakdown of each level). Yeates (2018) says, "such learning requires that their choices are achievable. Empowering animals to make—and achieve—better choices could lead to them making better choices than either party can make otherwise" (p. 183). Teaching skills through these methods is a way to provide information the dog needs to make informed decisions and reduces the amount of fear, stress, or frustration they are likely to experience otherwise.

In Chapter 6 of this book, I outline a canine-indexed definition of consent that purposefully constructs consent from a dog's perspective. But let's start thinking about simple ways we can increase agency and the opportunities our dogs have to make more opportune and desirable choices in their everyday lives before we get there. These might look like:

1. Creating spaces ("safe zones") a dog can access in the home where they will not ever be touched or disturbed by people.
2. Teaching them cooperative care and the use of "consent behaviours" (discussed in Chapter 4).
3. Offering choices between a variety of chew items and treats.
4. Free offering of food (rationed if needed) in a variety of desirable ways (bowl, scatter, hidden, scavenge, puzzle toys, etc).
5. Free access to toys.
6. Freedom to choose pace and stop/start points, and the route on walks.
7. Teaching the skills needed to have as much off leash time as possible and/or using a long lead to allow greater freedom of movement.
8. Providing free time for sniffing and scavenging.
9. Allowing dogs to choose equipment (for example, type of harness or collar they prefer).
10. Asking them if they would like to engage in cuddling, grooming, etc., or simply walk away.
11. Noting the things that your dog needs but is less inclined to assent to on their own (e.g., nail trims) and work on introducing it slowly and incrementally, allowing them to choose to participate at each small step (or not) until they are comfortable with the end behaviour (more about shaping in Chapter 4).
12. Allowing free access to indoor and outdoor space as much as possible and as much as is safe for that individual and environment; this might be a dog door, leaving doors open in warm weather, or asking frequently if they want to go out or in.
13. Providing the freedom to meet new people (or not) at their own pace without risk of unwanted groping.
14. Creating social opportunities for socially inclined dogs to make new friends and play freely if they choose. Dogs who struggle socially might do well having a few friends that they are able to interact with on planned occasions.
15. Free choice of activities.
16. Consent checks: pausing while petting/interacting to observe their yes/no signals Communicating the option to walk away or end the interaction.

3.6 CONCLUSION

Our companion dogs are significantly dependent on us, and this leaves them vulnerable to exploitation and violations of any supposition of agency they may experience. Even in the most seemingly benign ways, we impose our needs and wants onto our canine partners more often than we consider their subjective needs and desires. Most companion dogs live lives that are far from autonomous, many have little ability to choose how they spend their time or make decisions about their own lives, and their expressions of interest often go ignored or overruled. It is important to make clear that believing dogs have rights does not mean believing dogs should necessarily be treated the same as humans or that every interaction between a human and a dog is morally suspect. Offering choices also doesn't mean always having free choice (Hazel, 2021), but at the very least dogs should have free choice in situations that don't compromise safety and be taught skills that support better choices in future situations. All members of society have rules that need to be followed, and it is our responsibility to teach social rules in a humane and compassionate way that anyone deserves. And as a part of this, it includes providing salient choices and the ability to consent or withdraw that consent at any time.

NOTES

1. Stimulus is something in the environment that creates a physical or behavioural reaction, and discriminative stimuli is a stimulus that predicts reinforcement.
2. An event, situation, or object that evokes avoidant or escape behaviour. In other words, something that the dog finds unpleasant.

REFERENCES

Amdam, G. V., & Hovland, A. L. (2011). Measuring animal preferences and choice behavior. *Nature Education Knowledge, 3*(10), 74.

Byrne, R. (1995). *The thinking ape: The evolutionary origins of intelligence.* Oxford University Press.

Catania, A. C. (1975). Freedom and knowledge: An experimental analysis of preference in pigeons. *Journal of the Experimental Analysis of Behavior, 24,* 89–106.

Catania, A. C., & Sagvolden, T. (1980). Preference for free choice over forced choice in pigeons. *Journal of the Experimental Analysis of Behavior, 34,* 77–86.

Corrieri, L., Adda, M., Miklósi, Á., & Kubinyi, E. (2018). Companion and free-ranging Bali dogs: Environmental links with personality traits in an endemic dog population of South East Asia. *PLoS One, 13*(6), e0197354.

Donaldson, S., & Kymlicka, W. (2011). *Zoopolis: A political theory of animal rights.* Oxford University Press.

Freidman, S. G. (2020). Why animals need trainers who adhere to the least intrusive principle: Improving animal welfare and honing trainers' skills. *IAABC Journal.* https://journal.iaabcfoundation.org/why-animals-need-trainers-who-adhere-to-the-least-intrusive-principle-improving-animal-welfare-and-honing-trainers-skills/

Griss, S., Riemer, S., Warembourg, C., Sousa, F. M., Wera, E., Berger-Gonzalez, M., Alvarez, D., Malo Bulu, P., López Hernández, A., Roquel, P., & Dürr, S. (2021). If they could choose: How would dogs spend their days? Activity patterns in four populations of domestic dogs. *Applied Animal Behaviour Science, 243,* 105449.

Hazel, S. (2021). Understanding emotions. *Companion Animals New Zealand*. www.companionanimals.nz/videos21

Jones, E. (2021). What can "streeties" teach us about companion dogs? *The IAABC Foundation Journal, 19.*

Kirkden, R. D., & Pajor, E. A. (2006). Using preference, motivation and aversion test to ask scientific questions about animals' feelings. *Applied Animal Behaviour Science, 100*, 29–47.

Lagisz, M., Zidar, J., Nakagawa, S., Neville, V., Sorato, E., Paul, E. S., & Løvlie, H. (2020). Optimism, pessimism and judgement bias in animals: A systematic review and meta-analysis. *Neuroscience & Biobehavioral Reviews, 18*, 3–17.

Marshall-Pescini, S., Passalacqua, C., Petrazzini, M. E. M., Valsecchi, P., & Prato-Previde, E. (2012). Do dogs (*Canis lupus familiaris*) make counterproductive choices because they are sensitive to human ostensive cues? *PLoS One, 7*(4), e35437.

Martin, T. L., Yu, C. T., Martin, G. L., & Fazzio, D. (2006). On choice, preference, and preference for choice. *The Behavior Analyst Today, 7*(2), 234–241.

McMillan, F. (2019). The mental health and well-being benefits of personal control in animals. In F. D. McMillan (Ed.), *Mental health and well-being in animals* (2nd ed., p. 67). CABI.

Milligan, T. (2017). The ethics of animal training. In C. Overall (Ed.), *Pets and people: The ethics of our relationships with companion animals* (pp. 203–217). Oxford University Press.

Nussbaum, M. C. (2023). *Justice for animals: Our collective responsibility*. Simon and Schuster.

Paul, M., Majumder, S. S., Sau, S., Nandi, A. K., & Bhadra, A. (2016). High early life mortality in free-ranging dogs is largely influenced by humans. *Scientific Reports, 6*, 19641.

Pierce, J., & Bekoff, M. (2021). *A dog's world: Imagining the lives of dogs in a world without humans*. Princeton University Press.

Prato-Previde, E., Marshall-Pescini, S., & Valsecchi, P. (2008). Is your choice my choice? The owners' effect on pet dogs' (*Canis lupus familiaris*) performance in a food choice task. *Animal Cognition, 11*, 167–174.

Range, F., Heucke, S. L., Gruber, C., Konz, A., Huber, L., & Virányi, Z. (2009). The effect of ostensive cues on dogs' performance in a manipulative social learning task. *Applied Animal Behaviour Science, 120*(3–4), 170–178.

von Uexkull, J. (1957). A stroll through the worlds of animals and men: A picture book of invisible worlds. In C. H. Schiller (Ed. & Trans.), *Instinctive behavior: The development of a modern concept* (pp. 5–80). New York: International Universities Press.

Wehmeyer, M. L., & Schwartz, M. (1998). The relationship between self-determination and quality of life for adults with mental retardation. *Education and Training in Mental Retardation and Developmental Disabilities, 33*, 3–12.

Yeates, J. W. (2018). Why keep a dog and bark yourself? Making choices for non-human animals. *Journal of Applied Philosophy, 35*(1), 168–185.

CHAPTER 4

TEACHING CONSENT AND COOPERATIVE CARE

4.1 INTRODUCTION

The New Zealand Health Education Association says that "Learning about consent includes the development of knowledge, skills and understanding that are transferable to other . . . contexts where consideration of consent and other situations involving respectful and effective communication feature" (Robertson & Dixon, 2022, para. 3). I have heard the argument that cooperative care isn't consensual because it's simply a dog performing a learned behaviour. But first, I would argue, this is far too reductionist for the realities of our dogs' extraordinary capabilities and experiences. Secondly, I would argue that we are all animals who are acting on learned behaviours, behaviours that form a function. Dogs, too, are able to develop their knowledge and understanding about handling for medical care, grooming, or other types of touching/handling, with assistance from their caregivers. They are able to learn skills that help them to communicate if we provide them with such opportunities to garner the appropriate systems. The underlying aspects of consent are features of any healthy relationship, whether between humans or between humans and dogs, and rely upon mutual respect and a connection to the physical, social, and mental dimensions of well-being. Therefore, consent has implications to our connectedness across species boundaries: The quality of our relationships with our dogs, trust, and mutual respect. This includes the emotions they feel when relating to humans, their level of comfort and security, how they feel about themselves, and what they value, such as their physical safety and bodily autonomy.

Largely, teaching dogs about consent means teaching them that they have choices and that those choices will be respected. Cooperative care can be an example of just that. Cooperative care—a specific technique designed to empower dogs to have control over what happens during a specified touch-based interaction—is a way to teach dogs the skills that allow them to consent and withdraw consent at any point they choose when taught correctly (Jones, 2023). As well as the process of teaching them about consent, it is a way to introduce new handling procedures without stress so that consent is more likely to occur and continue, allowing us to attend to their physical health and wellness while concurrently addressing their emotional wellness. This makes interactions that involve specific hands-on procedures predictable

DOI: 10.1201/9781003361459-4

and safe, reducing or eliminating stress or anxiety related to arbitrary touching. It can also provide dogs with the information they need to make adequately informed decisions about the touching that happens to them, even if they don't understand the complexities of why it might be important for it to occur. It is a protocol founded on trust between dogs and humans and forms a clear reciprocal communicative system that seeks to reduce fear or stress during otherwise potentially stressful handling. So while much of what I have discussed in Chapters 1–3 talks about intrinsic body language systems that can be used to indicate consent in most everyday interactions, cooperative care is taking it another step further by teaching dogs about how to consent and how to retract consent.

Though cooperative care is the primary focus of this chapter, I will also explore some other ways we can introduce consent and/or, at the very least, provide more choices when teaching new behaviours or modifying behavioural patterns. Cooperative care, and some of the other protocols we will discuss, challenge traditional beliefs conflated with forceful and remedial methods of handling that are woven into the fabric of our human-centric society and traditional veterinary care system, often propagated further by the misinformation and misconceptions that are freely accessible to anyone with internet access. Changing social norms, those dictated by human exceptionalist paradigms, can be challenging to vanquish and certainly directly influence how we choose to interact with dogs. However, it is necessary for our exchanges with dogs to become more ethical and compassionate and to evolve into a more sophisticated approach for the enhancement of their well-being.

4.2 TEACHING CONSENT THROUGH COOPERATIVE CARE

Cooperative care is not about compliance, though it almost always leads to a dog who is willing to participate, at least to the degree for which they're prepared. The main goal of cooperative care is to allow dogs to choose to actively contribute to their own care and to give them control over what happens to them. Though cooperative care is often used in the realm of veterinary practice, it is actually about consent in *all* interactions that involve handling, both in the veterinary clinic and at home. The foundation skills learned in cooperative care should go beyond choosing to participate (which is also great because that is what we want to happen when we need to handle our dogs for their health and physical wellness), but also include teaching them they have the ability to consent or withdraw that consent at any point they choose. That means they know that it's not only okay to say no, but to trust that their decision will be respected. I do want to mention that I believe the majority of cooperative care practices out there are not necessarily consensual and end up being negotiations rather than free choice. But more importantly, whether we teach/use learned consent behaviours, any procedural handing should be focused on reducing stress, increasing choice and autonomy, and include voluntary cooperation. In order for cooperative care to be consensual, dogs must have free choices that are moderately equal and tenable for both consenting or choosing to walk away.

Cooperative care first involves teaching the dog a consent-based behaviour using positive reinforcement (see Table 4.1 for definitions of categories of learning). For example, a chin rest is a common consent behaviour where resting the chin on an object signals the start of handling and lifting the chin indicates redaction of consent and all handling must pause. Other consent behaviours might include a sustained nose to object target, a particular position (example, lateral recumbency as a consent position for a lateral saphenous blood draw), or stationing on a low platform or mat. Teaching dogs to opt out (remove consent) is the second behaviour that should be taught. For example, if Juno is resting her chin to tell me she consents to being handled, she may lift her head to tell me she withdraws that consent. Both behaviours should be clear before we begin cooperative care. Once these behaviours are learned and performed with duration in absence of any additional handling, we can then teach them about various handling that might take place using conditioning and desensitisation techniques (Table 4.1) while providing them with **predictive cues**—cues used to specifically indicate what handling is to follow.

Conditioning and desensitisation protocols are used to slowly shape the consent behaviour as new procedures are introduced (for example, medication administration or brushing). This involves slowly increasing the duration of the consent behaviour and adding in verbal cues that will correlate with what actions will follow (predictive cues). Once a consent behaviour is learned, it is easy and quick to transfer to new handling protocols. For example, I might teach Juno a sustained chin rest as the consent behaviour, where she fully understands that the chin rest is a "green light" for touching and lifting her chin is a "red light" indicator for touching to stop. With the addition of learned predictive cues, I can trim nails, apply ear treatment, or give vaccines, provided each of these specific handling protocols have been conditioned separately. In fact, I was able to remove stitches using a consent behaviour even though this is not something we have specifically practiced before. But because Juno has a strong knowledge of how to use her consent behaviour, and I have used predictive cues for touching her leg (where the stitches were located) and because using cooperative care in the past has built up a strong foundation of trust, it was seamless for her to transfer this knowledge to a slightly novel procedure very quickly.

The following is an example of the steps I used to teach a vaccine protocol to Juno using a chin rest consent behaviour (also see Figures 4.1 and 4.2).

Example: Teaching Vaccine Injection Using a Chin Rest Behaviour

1. Teach the chin rest and chin rise behaviours on their own. End each practice by using a terminal marker (using a clicker or the word "yes") following immediately with a reward to reinforce the behaviour. The marker indicates the correct behaviour and reinforcement is to follow (the marker acts as a secondary reinforcer).

2. Begin to add duration to the chin rest slowly and move at a pace with which your learner is comfortable. Mark the end of each duration with the marker and follow with a reinforcer.

3. Once the chin rest behaviour is clear to your learner and they are successfully able to sustain it for at least 10–20 seconds, begin to use instrumental conditioning to teach the various vaccine protocol steps that will take place while sustaining a chin rest. These suggested steps will only move as quickly as your learner can comfortably and confidently manage. Alternative options should be provided (toys, a bed, water, a second and third behaviour option, etc.), and sessions should remain short (~2–3 minutes) and positive for the learner. Anytime that consent is withdrawn, all touching should stop and only start again when the learner chooses to offer a chin rest. Each step should be repeated until it's well understood by the learner, and each behaviour needs to be marked and reinforced when complete for clarity of the criteria.

   a. Add hand movements and work your way toward their neck (or shoulder). Work toward being able to place you hand onto their neck while they sustain a chin rest behaviour.

   b. As you pat your dog's neck, pair the action with the verbal predictive cue "pet." This will teach the dog the meaning of the predictive cue. The cue should precede touching.

   c. Once the "pet" action and the subsequent predictive cue are solid, follow the action with pinching the skin gently and pairing the verbal predictive cue "pinch."

   d. Once the "pet" and "pinch" actions are solid, use a finger to poke, mimicking an injection and pair with the verbal cue "poke."

   e. Chain the pat, pinch, poke together sequentially.

   f. Replace poking with a finger with an empty, capped syringe followed in due time with a blunt needle. Remember that the predictive cue should precede each action, and it should be the same sequence every time: "pet," "pinch," "poke."

   g. Follow the preceding steps, you can repeat with the vaccine in the syringe without injecting it. Remember that a dog's olfactory senses are much more acute than our own, and this should be considered during this conditioning and desensitisation process.

   h. Repeat these steps in the environment where you will be giving the vaccines and with the person who will be administering them. When starting in a new environment and/or when introducing a new person, begin with an easy step (example, just "pet") and work up to the full sequence. This allows you to gauge your dog's body language and not accidentally push them beyond their level of tolerance. The experience should never be aversive to the dog.

When ready, you can do the full vaccine. It is advisable to use a numbing cream on the area so this may need to be conditioned separately in advance. Most countries require vaccines to be administered by the veterinarian so this should be practiced in advance ideally with the doctor who will be doing the administration.

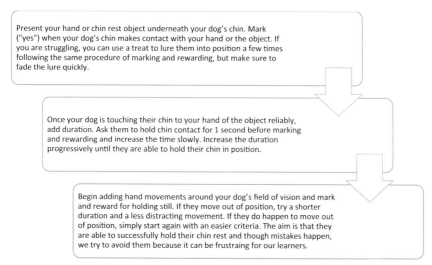

Present your hand or chin rest object underneath your dog's chin. Mark ("yes") when your dog's chin makes contact with your hand or the object. If you are struggling, you can use a treat to lure them into position a few times following the same procedure of marking and rewarding, but make sure to fade the lure quickly.

Once your dog is touching their chin to your hand of the object reliably, add duration. Ask them to hold chin contact for 1 second before marking and rewarding and increase the time slowly. Increase the duration progressively until they are able to hold their chin in position.

Begin adding hand movements around your dog's field of vision and mark and reward for holding still. If they move out of position, try a shorter duration and a less distracting movement. If they do happen to move out of position, simply start again with an easier criteria. The aim is that they are able to successfully hold their chin rest and though mistakes happen, we try to avoid them because it can be frustrating for our learners.

Figure 4.1 **Process of teaching a chin rest using positive reinforcement.**

Cooperative care, when taught correctly, allows the dog to understand that they have a choice, but there is also an onus for us to understand the dimensions at play to truly make cooperative care an act of consent. During any cooperative care practice or procedure, the following needs to be evaluated and considered on a continual basis.

4.2.1   Stress Assessment

If at any point during any of the steps listed in the example the dog lifts their head out of the chin rest (or presents an otherwise learned consent withdrawal behaviour), all handling must cease. Assess the dog's emotional state, and if they are displaying "no" signals (see Chapter 2) it may be best to end the session and try again with potentially lowered criteria in the future. During these teaching sessions, the dog is learning that they have control over what is happening to them, and this control to end the session needs to be reinforced.

4.2.2   Reinforcement Contingencies and Criteria

The dog should feel safe enough to opt out at any stage, even before any touching has commenced. Only reinforcing the consent behaviour might lead to coercion (they feel they need to continue to consent because they want the food reward but are not genuinely consenting to the interaction). Dogs may stay in uncomfortable situations if there are competing reinforcers (and a stronger motivator present). The Matching Law suggests that the more likely behaviour the dog will choose when presented with options is the one that has a stronger history of reinforcement, which may bias our dogs' decision.

To emphasise this potentially false consent behaviour, I will draw on the example of dog Lily and my human friend, Jenna. Lily was a keen participant and offered her chin rest without hesitation. But she didn't waver from this offered behaviour, even when handling made her uncomfortable. My friend soon noticed that Lily was holding her chin rest "consent" behaviour but growling as imposing hands approached her body. What went wrong? Lily either a) was unclear that lifting her head from the chin rest was a way to remove her consent, or b) she knew the criteria to remove her consent was to lift her head but was conflicted because she wanted the food reward for holding her chin rest position. Therefore, the chin rest was not acting as a consent behaviour at all.

The choice of reinforcer for consenting and for removing consent may differ depending on the individual's preferences. It might be equal for both consent and consent withdrawal (food or toy), which is often what I do for Juno. I reinforce her with food regardless of if she chooses to continue to consent or if she chooses to withdraw consent and this has not affected her choice to consent in most situations. This is because the steps we have taken have proven to provide a feeling of safety for her and are predictably a fun way for her to engage with me. Proof is in the pudding here. When I grab a syringe or the nail clippers, she tells me "yes" enthusiastically and with great fervor because it is clearly something she enjoys.

There is also an inherent reinforcement that occurs when a dog withdraws consent and the handling stops if a dog is truly feeling uncomfortable. For example, if Juno is feeling stressed while being handled and chooses to lift her chin (withdraw consent), then the stress of being touched is removed (which is negative reinforcement), though the aim is to move at a pace slow enough that she is never feeling worried. Additionally, the power of having control over what happens to her own body in that moment of stress or uncertainty is also positively reinforcing. Dr. Susan Friedman (2020) suggests that *choice is in fact a primary reinforcer* that will serve to strengthen a behaviour over time. Therefore, the dog may choose option a or option b based on two different motivating factors.

However, we have to ensure that criteria are clear to our dogs. If the dog is relaxed but simply struggling with understanding the chin rest behaviour, it is likely that the criteria have become muddied, and they haven't fully learned how to utilise their consent behaviour to control the situation. It may be advisable to return to adding duration to the chin rest behaviour in absence of any cooperative care handling to ensure the behaviour is understood. Because the aim of cooperative care is about predictability and clarity, if the criteria is muddled or learning is incomplete, this may cause an extra layer of stress or frustration. To reduce any potential stress and to increase clarity of criteria, it is important to move at the pace the learner sets and to remain constantly vigilant to their emotional state.

4.2.3 Providing Salient Options

Choice is what cooperative care is about. Consent is only available to use if there are salient choices to be had. Therefore, having toys, optional human-free space, water, and chew items located within range so the dog has the option to engage with something else if they are not comfortable with handling. There should be options to leave (i.e., they are not locked in a room, backed into a corner, etc.). I also teach Juno that she has the option to do a different behaviour (one that doesn't result in handling or touching at all). For her, this is normally a nose target because she finds this one easy and fun, and she seems to genuinely enjoy poking me with her nose. The key is that we need to be sure we are not leveraging access to choices in order to reach our own goals.

4.2.4 Frequent Consent Check-Ins

It is important to check in with the dog frequently to see if the interaction is still consensual. To test this, toss treats away to reset their position. If they return and offer their consent behaviour, this is our indication to continue. If not, a break or session end is needed.

4.2.5 Only Ask if You Can Accept "No" for an Answer

If the required procedure is life altering and/or is time sensitive or if the procedure is something they haven't yet learned, don't ask for their consent. The goal of cooperative care is about maintaining the trust and communication in dog–human interactions. If there is a risk that Juno will say no, and I then restrain her and do it anyway, I have put her (and myself) in an unjust situation and delegitimised her request to withdraw consent. In such a situation, it is better to take the least intrusive and least stressful approach, which often will be sedation or potentially highly valued distractions. For Juno, this does not happen often as most procedures are well learned and, in most cases, we can postpone if needed.

Notice I said most cases and not *all*. Back in mid 2022, Juno became very ill. She needed a series of tests that required more invasive handling than what she could comfortably consent to or what we were prepared for. Therefore, pre-medical testing and right up to pre-surgery (she ended up having a splenectomy), we sedated her as quickly and efficiently as possible in an environment in which she felt safe and comfortable. I still used predictive cues ("poke") and I still used a preconditioned "hug" cue for a light restraint and thus was able to provide her with information even though I didn't ask her for her consent. We were then able to reverse sedation just as we prepared to leave the clinic, and I did most of the handling at home (for example, dressing changes, intravenous catheter removal, analgesic injections, oral medications) using consent behaviours. This means that we cut out the stress of being in waiting areas and exam rooms, restraining, touching her without permission while she was fully aware and potentially nonconsenting, and thus, we were able to do most of the handling without stress.

Figure 4.2 Juno demonstrating her chin rest behaviour while practicing vaccination protocol with an empty syringe and blunt needle.
Source: Photo credit: Erin Jones.

Table 4.1 Categories within Learning Science. Includes Skinnerian Operant Conditioning, Pavlovian Classical Conditioning, Non-Associative Learning and Other Influences on the Learning Processes such as Observational Learning and Socialisation

TYPE	CATEGORY	PROCESS	DESCRIPTION
Associative Learning	Operant (instrumental) Conditioning	Positive Reinforcement	Adding a reinforcer to strengthen a behaviour.
Associative Learning	Operant Conditioning	Negative Reinforcement	Removing an unpleasant stimulus to strengthen a behaviour.
Associative Learning	Operant Conditioning	Positive Punishment	Adding an aversive stimulus to weaken a behaviour.
Associative Learning	Operant Conditioning	Negative Punishment	Removing a pleasant stimulus to weaken a behaviour.

Table 4.1 (Continued)

TYPE	CATEGORY	PROCESS	DESCRIPTION
Associative Learning	Classical Conditioning	Associations	Pairing a neutral stimulus with a primary reinforcer or aversive stimulus to create a good or bad association with the neutral stimulus.
Associative Learning	Classical Conditioning	Extinction (and extinction burst)	No longer reinforcing a once reinforced behaviour. This will weaken the behaviour over time. This behaviour is likely to resurface after extinction (extinction burst).
Associative Learning	Classical Conditioning	Systematic Desensitisation	Exposure therapy aimed at removing the fear response with a relaxation response with slow, gradual exposure to the stimulus.
Non-Associative Learning	Habituation	Habituation	Becoming acclimated to a non-threatening stimulus.
Non-Associative Learning	Sensitisation	Sensitisation	Becoming more aware and fearful or anxious of a stimulus and generalising to a similar stimulus in the environment.
Observational Learning			Learning through observing conspecifics.
Socialisation	All Categories	All Processes	The process of learning key life skills. This includes all the processes of learning during exposure to different stimuli in their environment.

Critical Corner: Always ask yourself:

1. How might the rate of reinforcement and level of practice influence cooperation?
2. Has the exit strategy also been practiced and reinforced?
3. How can we be sure we are not leveraging access to choices to reach our own goals and can we ensure we are not inadvertently stripping our dog of salient choices?

4.3   WHY COOPERATIVE CARE: EMPOWERMENT

Physical care is important for dogs' welfare and their well-being, though studies have shown that veterinary care may have a significant emotional impact on many dogs, manifesting in fear-related behaviours stemming from fight or flight response patterns (Figure 4.3). For example, Stanford (1981) found that 60% of dogs exhibit trepidation and deference/appeasing signals when visiting the vet, and 18% had a history of biting during veterinary visits. A study by Mariti et al. (2015) identified fear-related behaviours in 53% of dogs who entered the waiting room, and Döring et al. (2009) found that 50% of dogs were fearful to even walk into the examination room. In fact, fear-related behaviours, from mild to severe, can be observed across all of the different stages of a visit to the veterinary clinic, but are particularly evident during the examination process, especially those procedures involving physical touching/handling (Stellato et al., 2020, 2021).

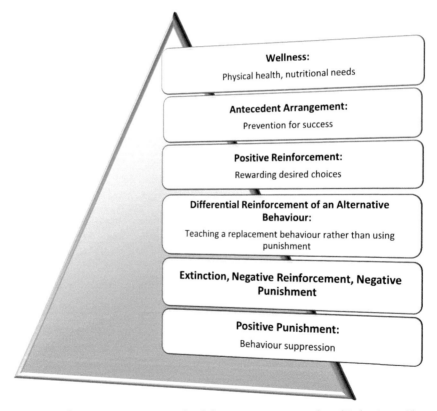

Figure 4.3 The systematic approach of the Humane Hierarchy of Behaviour Change (created by Dr. Susan Friedman, 2008).

There is an obvious and crucial concern for the physical well-being of dogs in veterinary care. In fact, the first step in psychologist and behaviourist, Dr. Susan Friedman's (2008) Humane Hierarchy[1] is Wellness (Figure 4.3) because physical and nutritional health needs to be assessed before attempting to create a behaviour intervention plan in order to rule in or out any potential underlying pain or medical concerns. This is a crucial beginning for the best laid behaviour plans to be effective and for the basic needs of any dog to be met, especially because there are strong correlations between physical health and behaviour. For example, Mills et al. (2020) found that approximately one third of referred behaviour cases (and up to 80%) involve some form of pain-related condition. These comorbidities include musculoskeletal, gastrointestinal, and dermatological conditions, which are commonly identified as significant to an animal's manifesting behavioural issues.

However, putting the emotional well-being at the forefront of medical care is essential for improving positive welfare as has been demonstrated. Susan Friedman (2020, para 4) suggests that behaviour is an "evolved tool to achieve functional outcomes to realizing that control over outcomes matters in the lives of all animals," and having a choice over control is part of what helps us create a better, more just relationship with dogs. Plenty of multispecies scholarly evidence recognises a direct correlation between control and positive welfare (for example, see Friedman, 2005; Leotti et al., 2010), and as such, control (over oneself) has been recently added as a pillar of positive animal welfare according to Mellor's (2020) Five Domains model). Why is this important? Initially, animal welfare focused on reducing or eliminating negative impacts on welfare until Mellor's more recent revamp of welfare standards. This addition of **positive welfare**, which in part includes "a sense of control," is part of that focus. To further address the necessity of having a sense of control, Friedman (2020) also suggests that control can be considered a primary reinforcer—something that is inherently valuable and rewarding that strengthens a behaviour over time as I discuss earlier in my example of teaching consent behaviours. Controlling our outcomes is the adaptive function of behaviour; we need to control outcomes to not only survive but to flourish.

Cooperative care can *empower* individuals to contribute to a reciprocal dialogue that ultimately affects their emotional wellness and right to choose. But it is more than just empowerment. It is also about feelings of safety for both humans as well as dogs. It doesn't just provide the dog with a choice to cooperate, it delivers predictability and transparency to which they can base their decision to consent or not. When we remove force, or the threat of force, we eliminate the associated fear and anxiety—fear of the unknown (what is about to happen to their bodies) but also fear of the known (what happened in similar instances). It also inspires trust, the foundation of any successful relationship.

Distress can manifest in many ways for a dog. Some of these pose a risk to the veterinary team, others may not. You may have heard of the fight, flight, or freeze response, when a dog's brain essentially switches into survival mode; it is a response of the sympathetic nervous system (SNS) caused by a release of adrenalin and nor-adrenalin of which are completely involuntary (See Figure 4.4). These hormones create chemical messages within the brain to engage other organs vital for survival. This causes heart rate and breathing rate to increase, blood pressure to rise, and other parts of the body to become inhibited, like digestion. Often fear-based behaviours manifest as a part of the dog's defensive mechanism (example, a bite), which occur more often and more quickly when an animal is under distress. When there is no opportunity to flee, whether this is because of restraint or confinement (literal or implied), including things like leashes and closed doors/small spaces, fear of reprimand, and lack of escape routes, self preservation may lead to a fight response. But just because aggressive behaviours may be more damaging to humans, other fear responses should be taken just as seriously. For example, the lack of behaviour doesn't mean that the internal state of an animal is any different to that of an animal who behaves defensively. Freezing may equally indicate internal distress in an

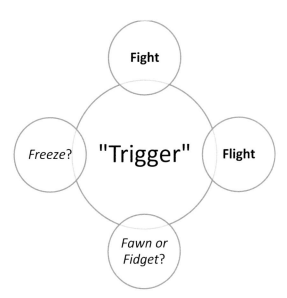

Figure 4.4 The responses to a perceived threat may vary depending on individual and circumstances. Fight responses are often aggressive behaviours used to threaten including snarling, snapping, and biting. Flight refers to the avoidance of the threat by moving away, hiding, cowering, tucking tail, or lowering their body to the ground. Freezing, fawning, and fidgeting all contain a question mark by them because the expansion to include other various responses to stress (which include sometimes displacement behaviours, stress signals, and/or calming signals) may vary and is not fully known. This has led to many researchers to call the set of responses an "acute stress response" (Bracha, 2004).

animal, and this is frequently mistaken for an animal who is compliant or "obedient." There are arrays of various responses we may see from a dog who is anxious or fearful, which have been addressed in Chapter 2.

4.4 CONSENT IN LEARNING PROTOCOLS

The goal of teaching should be about providing skills that help dogs to navigate their environment with the most amount of success and provide value to their lives. But the methods we choose should also reflect consensual interactions and maximise agency. Traditional methods of teaching dogs revolved around the praxis of a dominant-submissive relationship framework (Włodarczyk, 2018) and therefore involved punishment as a form of behavioural control. These methods are never consensual because they use, at the very least, forms of coercion, threat, and intimidation. Though research has demonstrated that the use of aversives in training (particularly positive punishers but also some negative reinforcers) have fallout effects, including an increase of aggressive behaviours, apathy, and escape behaviour (example, Heron et al., 2099; Hiby et al., 2004; Roll & Unshelm, 1997; Blackwell et al., 2008; Skinner, 1957; Ziv, 2017), it seems deeply entrenched and remains common practice. Additionally, dogs taught using aversives to suppress behaviour can experience greater levels of stress and thus may behave more aggressively toward people (Casey et al., 2014; Deldalle & Gaunet, 2014; Ziv, 2017). Dogs who are taught with aversives also demonstrate a lower quality relationship with their human caregivers, demonstrated by showing less affection and less gaze direction (de Castro et al., 2020; Rooney & Cowan, 2011). However, misunderstandings lead to a reframing of particular punishment techniques as "gentle" or "not painful" which, by definition of a punisher, means they would in fact be useless. In an earlier study (Jones, 2022) I found that rather than focusing on the pain aspect of physical punishment, people tend to shift the narrative to "submission" and "obedience" as a motive for physical control or aversive training tools, with methods or techniques that "don't really hurt," but instead relying on fear (threat), intimidation, or other annoyances to aim to suppress behaviour. As I have discussed in Chapter 2, aggressive behaviours are most often last resort behaviours in many instances where a dog is feeling fearful or stressed, and the suppression of behaviour does nothing to address the internal emotional state the dog may be experiencing.

People often justify their forceful or coercive actions, based partially on the objectification of dogs as property and their view that dogs have a subordinate role to humans even when they are considered part of the family (Greenebaum, 2010; Jones, 2022). The intrinsic power relations people have over dogs is a deliberately difficult truth; however, it is important to acknowledge if we are to support any kind of good life for dogs. Of course, there are certain responsibilities we carry as accountable caregivers: we have responsibilities for our dogs' lives from beginning to end, and this means that we will have an immense contributory influence on the quality of their lives.

Irvine (2004) suggests that "relationships between humans and animals have depended on how a given society defines animals and what it means to associate with them," arguing that "what we currently know about animals demands wrestling with the moral implications of keeping them as pets" (Irvine, 2004, p. 5). This view, shared by other animal ethicists (for example, Pierce, 2016), could serve as a helpful engagement to strategise difficulties that are often overlooked, specifically problems that point beyond welfare issues and toward other normative perceptions of our relationship with dogs. Benz-Schwarzburg et al. (2020, p. 13) says,

> In order to treat dogs in the way that morality requires of us, it is paramount that we bear in mind the spectrum of positive duties that this relationship engenders, including the duty to live up to the trust that dogs place in us.

Part of this includes a responsibility to teach about consent, ask for consent, and respect the withdrawal of consent both in cooperative care and within other daily interactions with dogs, including the way we teach them skills.

Despite our growing knowledge of how such forceful and coercive techniques impact well-being, teaching consent behaviours for cooperative care (e.g., grooming, bathing, nail trims, veterinary exams, or blood draws) is only just becoming evident in the mainstream discourse of dog teachers and educators. Cooperative care and other such dog-focused teaching/learning paradigms can provide information through experience and predictive cues, predictability through routine, and teaching through gain (positive reinforcement) rather than avoidance (negative reinforcement or positive punishment). Positive reinforcement is built on the foundation of providing information and building communication through a reward-based system. Though this type of transaction of information may not fit the human definition of consent, it is an ideal communication tool allowing for choice and autonomy in situations that normally utilise force and restraint, and there is no reason why learned behaviours cannot be considered as a species-specific consent-based interaction. However, this movement toward consensual interactions with dogs is in its infancy and the problems people may face involve its practical application and the conjecture of implementing these systems fairly. In particular, its application in industries (for example, veterinary clinics) where these practices are not (yet) commonplace can be particularly challenging, especially for the regular dog caregiver not involved professionally in the veterinary or behaviour industry.

Consent and agency can be applied in a variety of ways when teaching dogs to navigate this human-centric environment if we shift our focus from training dogs to perform or obey to instead think about teaching dogs the skills needed to live harmoniously within human confines. In fact, teaching functional cues/skills and concepts to enhance their abilities and happiness can offer needed guidance and information for dogs to live successfully and with less stress. Therefore, I want to emphasise that some skill learning is essential for well-being, and some guidance is valuable, particularly in situations where dogs look to us for reassurance, comfort, or direction to better navigate novel situations. The skills I choose to teach may be based on the individual's needs (areas where they may need extra support) but generally include recall (come to me), wait (stop moving forward), stay (stay in one spot), with me (walk beside me), "relax" (free to wander), and behind (move behind me if you're feeling unsafe by an approaching stranger). These will allow my dog learner the most amount of freedom while maintaining safety and following social rules. Other learning strategies might include specifically targeting areas of concern, such as if my dog learner struggles with fear or anxiety around other dogs or people and will involve methods targeting at teaching a replacement behaviour (skill) that will provide them with feelings of safety and predictability where they have maximum control over their bodies and space.

Earlier I mentioned Friedman's Humane Hierarchy of behaviour modification (Figure 4.3). This framework provides a systematic approach to teaching dogs while maximising their ability to choose and providing them with information to learn new skills rather than suppress their behaviour. The paradigm denotes using the least intrusive or invasive approaches in the following ways:

1. Physical and Nutritional Wellness: It starts with the dog's physical and mental wellness, such as ruling in or out health issues or pain and ensuring that the dog is living a robust, enriched life and their species, breed, and individual needs are met and exceeded. This starting point should address any distant antecedents that may affect behaviour and motivation.
2. Antecedent Arrangement: The Humane Hierarchy uses antecedent arrangements to set the dog up for success and provide choices that are both realistic to the context of the environment and to outcome success (aimed at reducing negative outcomes and increasing positive and effective outcomes). Though the behaviour we may be looking to prevent is often only undesired by human standards, behaviour can also be a measure of emotional valance or can become a form of stress (for example, mail delivery person comes to the door, dog barks, caregiver yells at dog which the dog finds stressful). Therefore, eliminating the need to bark at the postal worker by switching to a PO box has reduced the undesired behaviour by eliminating the environmental stimulus that causes stress and subsequent stress response, either innately (absence of mail delivery person eliminates the stress and response) or indirectly via the removal of potentially negative consequences (the dog is no longer punished for barking when the postal worker visits the house). We may be eliminating an undesirable choice, but we are also eliminating the presentation of that choice to begin with.
3. Positive Reinforcement (R+): R+ simply means that we are adding a reward to strengthen behaviours. For example, when we first begin to teach a dog to do a chin rest to eventually use as a consent behaviour, we could strengthen that behaviour by reinforcing the act of chin resting. Firstly, this strengthens the behaviour we want to see repeated—the skill that will set them up for

success. The Matching Law (Cooper et al., 2007) suggests that the "allocation of responses to choices are distributed in proportions that match the rates of reinforcement received from each choice alternative." This means that behaviour goes where the reinforcement flows! Secondly, it provides our dogs with the information they need as part of a system of communication. If I am teaching a new skill, one way I can tell the dog that they just gave the right answer to a problem is through using reinforcement—it's about providing positive feedback. It also means we are aware of the individual's preferences and choices. Understanding what Juno values may help me to provide her with the best possible consequences for her behaviours. I would add the disclaimer that I would never withhold a reinforcer in other situations just to leverage what I want. Juno gets lots of treats for lots of different behaviours, and she gets treats freely as well. That goes for praise, love, attention, access, toys, and all the other things that bring her enough joy that they make successful reinforcers.

4. Differential Reinforcement: The Humane Hierarchy then moves into using differential reinforcement. This is where we would reinforce an alternative behaviour choice as a behaviour change approach. Now, to be fair, the behaviour would have to be of an equal motivational value, and the consequence would have to prove to be equal to or better than the behaviour we are aiming to eliminate. I can't expect Juno to sit motionless when my husband returns home from work at the end of the day when what she would rather do is jump up all over him, especially if the reward I offer her for doing so is a pat on the head. I could, however, likely get her to engage in a game of tug when he comes in the door. I could also ask her to go to her bed and give her very favourite chew (which is extremely valuable to her), and she may then choose that new behaviour over jumping excitedly because the consequence is so valuable to her. This would have to be appreciably consistent to be most effective for a behaviour to change. There are various types of differential reinforcement (for example, reinforcement of incompatible behaviours or alternative behaviours; or lower or higher rates) and they can be a very effective alternative to punishing a behaviour we are trying to stop, and it also provides information about what we would like our dog to be doing in a particular context.

5. The rest of the Humane Hierarchy (extinction, negative reinforcement, negative punishment, and positive punishment) outlines some additional strategies that I won't get into here. However, the concept of using these guiding principles is that all of the least intrusive methods must be utilised and exhausted before considering steps that appear further down the hierarchy. If used correctly, there is no real reason that much more might be considered, and we should only be doing the minimal amount needed to help our dog to understand what is needed of them behaviourally to foster a successful outcome. Though do keep in mind that aversive consequences are inherent

in the environment as are things that will be negatively reinforcing for your dog. Remember earlier when I said that if the dog is feeling stressed and they recant their consent by lifting their head out of the chin rest position, they are being negatively reinforced? This isn't something we have intentionally applied in the learning plan in order to teach our dog a specific cue (i.e., we are not intentionally aiming to cause them stress), but it might occur naturally. The Humane Hierarchy is about the intentional application of particular protocols that utilise these different forms of learning science.

Many established protocols, such as Grisha Stewarts' BAT 3.0,[2] are based on the supposition that providing the most amount of perceived autonomy and providing semi-structured, though salient choices, is beneficial for dogs learning to navigate their environment when they are anxious and fearful and is a wonderful protocol that considers agency and even consent in some situations. Some additional established protocols might include the following:

- Control Unleashed (Leslie McDevitt) is a set of game-based teaching strategies that focuses on helping dogs become more resilient and relaxed in various ways.
- Shaping (which would fall under an example of the use of positive reinforcement and/or DRO in the Humane Hierarchy). During shaping, the dog is rewarded for specific benchmark behaviours toward an end goal (successive approximations). For example, I recently taught Juno a match-to-sample task. She had to choose an object from several choices that visually matched the object in my hand. Obviously, if I simply held up an item and waited, she would have no idea what she was supposed to do. I broke it down into small steps that were easy and fun for her so she could learn the rules of the game. Any teaching interactions always come with the option to get up and leave at any time.
- The Bucket Game (Chirag Patel) developed this protocol that allows the dog to indicate their choice to proceed with handling (similarly to consent behaviours). Instead of stationing or offering a chin rest, it has to do with looking at the container of treats. Handling begins when they focus on the container, and if the dog turns their head away, the handling stops.
- See Mark and Reward Training (SMART) X 50 (Kathy Sdao) allows you to simply capture and reinforce ideal choices a dog makes. For example, when I first started walking with Juno, I rewarded her every time she chose to "check in" with me (focus her gaze on/toward me). I didn't ask her, lure her, or ever put a cue to it. I simply strengthened her desire to "check in" by making it fun and rewarding. It is a way to teach her to walk on a leash without pulling and to walk off leash and still be "with me" as she explores her environment.
- Bond-Based Choice Training (Jennifer Arnold) is based on a mutual relationship that aims to help the dog create a secure attachment to their caregiver.

This is certainly not an exhaustive list but gives you an idea of the kinds of programmes that are becoming a part of the way we teach dogs. Now, granted, any can be taught using food (or other reinforcement) as leverage, but these are set up intentionally to avoid coercion, and if executed skilfully, could be used as part of a consensual teaching programme. The skills gained are also important for teaching dogs about consent because a large part of learning about consent is learning that you have choices and that your choices matter. These games and protocols are ways that we can teach dogs that they can not only say no, but how to say no.

NOTES

1. The Humane Hierarchy is a framework developed by Dr. Susan Friedman that outlines a systematic process for behaviour modification. It starts from the least aversive or intrusive approaches with an aim of avoiding unnecessary or inhumane teaching practices.
2. BAT 3.0 is short for the latest version of Stewart's Behaviour Adjustment Training and is a systematic learning system used to empower and inspire confidence in dogs using current knowledge from across disciplines.

REFERENCES

Benz-Schwarzburg, J., Monsó, S., & Huber, L. (2020). How dogs perceive humans and how humans should treat their pet dogs: Linking cognition with ethics. *Frontiers in Psychology*, 3587.

Blackwell, E. J., Twells, C., Seawright, A., & Casey, R. A. (2008). The relationship between training methods and the occurrence of behavior problems, as reported by owners, in a population of domestic dogs. *Journal of Veterinary Behavior*, 3(5), 207–217.

Bracha, H. S. (2004). Freeze, flight, fight, fright, faint: Adaptationist perspectives on the acute stress response spectrum. *CNS Spectrums*, 9(9), 679–685.

Casey, R. A., Loftus, B., Bolster, C., Richards, G. J., & Blackwell, E. J. (2014). Human directed aggression in domestic dogs (*Canis familiaris*): Occurrence in different contexts and risk factors. *Applied Animal Behaviour Science*, 152, 52–63.

Cooper, J. O., Heron, T. E., & Heward, W. L. (2007). *Applied behavior analysis* (2nd ed.). Pearson.

de Castro, A. C. V., Fuchs, D., Morello, G. M., Pastur, S., de Sousa, L., & Olsson, I. A. S. (2020). Does training method matter? Evidence for the negative impact of aversive-based methods on companion dog welfare. *PLoS One*, 15(12), e0225023.

Deldalle, S., & Gaunet, F. (2014). Effects of 2 training methods on stress-related behaviors of the dog (*Canis familiaris*) and on the dog–owner relationship. *Journal of Veterinary Behavior*, 9(2), 58–65.

Döring, D., Roscher, A., Scheipl, F., Küchenhoff, H., & Erhard, M. H. (2009). Fear-related behaviour of dogs in veterinary practice. *The Veterinary Journal*, 182(1), 38–43.

Friedman, S. G. (2005). He said, she said, science says. *Good Bird Magazine*, 1(1), 10–14.

Friedman, S. G. (2008). What's wrong with this picture? Effectiveness is not enough. *Good Bird Magazine*, 4-4. www.goodbirdinc.com

Friedman, S. G. (2020). Why animals need trainers who adhere to the least intrusive principle: Improving animal welfare and honing trainers' skills. *IAABC Journal.* https://journal.iaabcfoundation.org/why-animals-need-trainers-who-adhere-to-the-least-intrusive-principle-improving-animal-welfare-and-honing-trainers-skills/

Greenebaum, J. B. (2010). Training dogs and training humans: Symbolic interaction and dog training. *Anthrozoös, 23*(2), 129–141.

Hiby, E. F., Rooney, N. J., & Bradshaw, J. W. S. (2004). Dog training methods: Their use, effectiveness and interaction with behaviour and welfare. *Animal Welfare, 13*(1), 63–69.

Inglis, I. R., Forkman, B., & Lazarus, J. (1997). Free food or earned food? A review and fuzzy model of contrafreeloading. *Animal Behaviour, 53*(6), 1171–1191.

Irvine, L. (2004). A model of animal selfhood: Expanding interactionist possibilities. *Symbolic Interaction, 27*(1), 3–21.

Jones, E. (2023). Cooperative care for companion dogs: emotional health and wellness. *Companion Animal, 28*(7), 1–6.

Jones, E. E. A. (2022). *Silent conversations: The influence of human exceptionalism, dominance and power on behavioural expectations and canine consent in the dog-human relationship* [A thesis submitted in partial fulfilment of the requirements for the degree of Doctor of Philosophy in Human-Animal Studies, University of Canterbury]. https://libcat.canterbury.ac.nz/Record/3183157

Leotti, L. A., Iyengar, S. S., & Ochsner, K. N. (2010). Born to choose: The origins and value of the need for control. *Trends in Cognitive Science, 14*(10), 457–463.

Mariti, C., Raspanti, E., Zilocchi, M., Carlone, B., & Gazzano, A. (2015). The assessment of dog welfare in the waiting room of a veterinary clinic. *Animal Welfare, 24*(3), 299–305.

Mellor, D. J., Beausoleil, N. J., Littlewood, K. E., McLean, A. N., McGreevy, P. D., Jones, B., & Wilkins, C. (2020). The 2020 five domains model: Including human–animal interactions in assessments of animal welfare. *Animals, 10*(10), 1870.

Mills, D. S., Demontigny-Bédard, I., Gruen, M., Klinck, M. P., McPeake, K. J., Barcelos, A. M., Hewison, L., Van Haevermaet, H., Denenberg, S., Hauser, H., Koch, C., Ballantyne, K., Wilson, C., Mathkari, C. V., Pounder, J., Garcia, E., Darder, P., Fatjó, J., & Levine, E. (2020). Pain and problem behavior in cats and dogs. *Animals, 10*(2), 318.

Pierce, J. (2016). *Run, spot, run.* In run, spot, run. University of Chicago Press.

Robertson, J., & Dixon, R. (2022). *Teaching and learning about consent in health education: A resource to support learning in The New Zealand Curriculum.* NZHEA.

Roll, A., & Unshelm, J. (1997). Aggressive conflicts amongst dogs and factors affecting them. *Applied Animal Behaviour Science, 52*(3–4), 229–242.

Rooney, N. J., & Cowan, S. (2011). Training methods and owner–dog interactions: Links with dog behaviour and learning ability. *Applied Animal Behaviour Science, 132*(3–4), 169–177.

Skinner, B. F. (1957). The experimental analysis of behavior. *American Scientist, 45*(4), 343–371.

Stanford, T. L. (1981). Behavior of dogs entering a veterinary clinic. *Applied Animal Ethology, 7*(3), 271–279.

Stellato, A. C., Dewey, C. E., Widowski, T. M., & Niel, L. (2020). Evaluation of associations between owner presence and indicators of fear in dogs during routine veterinary examinations. *Journal of the American Veterinary Medical Association, 257*(10), 1031–1040.

Stellato, A. C., Flint, H. E., Dewey, C. E., Widowski, T. M., & Niel, L. (2021). Risk-factors associated with veterinary-related fear and aggression in owned domestic dogs. *Applied Animal Behaviour Science*, *241*, 105374.

Włodarczyk, J. (2018). *Genealogy of obedience: Reading North American pet dog training literature, 1850s–2000s*. Brill.

Ziv, G. (2017). The effects of using aversive training methods in dogs—A review. *Journal of Veterinary Behavior: Clinical Applications and Research*, *19*, 50–60.

CHAPTER 5

SUBSTITUTIVE CONSENT

EVALUATION FOR LONG-TERM HEALTH AND HAPPINESS

5.1   INTRODUCTION

Developing consent-based interactions with our dogs does not mean we will simply allow them to opt out of everything they don't want to do. It is about free choice in some situations and structured choices in others (for example, choices based on previously taught skills that involve consequences). And in a few situations, we may also need to make choices on their behalf and in their best interest—recall that in Chapter 1 we discussed substitutive consent, giving consent on behalf of a dependent made in their best overall personal interest (physically and emotionally). It is unlikely that dogs have the complex executive functioning that allow them to make fully informed decisions that plan for future wellness and long-term health and these decisions often involve major medical events that require such planning. We maintain certain responsibilities when we agree to care for dogs, particularly considering the many ways we have disempowered them through selective breeding, "training," and developing laws to restrict and control them to the point of creating life-long dependents. Recall that in Chapter 4 I briefly discuss "what-if-it's-impossible-to-ask-for-consent" scenarios, such as

- What if Juno says no, but it's something I feel will help her in the long term?
- What if she says no, but I do it anyway? What will that do to our relationship?
- What if it's something unpleasant that is needed but that she is likely to say no to, especially if she is largely uninformed and unprepared?
- What if it involves invasive procedures and invading her bodily autonomy but is necessary for improving her long-term wellness?

How do we navigate situations that require a deeper understanding of the best overall outcome when it's ultimately something *we* deem best *for* them? I am obviously not going to let Juno suffer in pain simply because she may not understand the long-term effects of surgery that will ultimately ease her suffering and allow her to live a long and pain-free life, simply because she says no to the sedation or is uninformed about what is about to happen to her body. The choices that people make, however, are often subjectively determined and there is really no entirely objective system that would ever be entirely complete to evaluate these situations. Someone else may have

DOI: 10.1201/9781003361459-5

instead chosen to euthanise Juno rather than pay thousands of dollars for her surgery. Yet another option someone may have chosen would be to maintain her health condition with pain-relieving drugs and limit her movements by using cages and barriers and short leashes until she died a natural death.

In this chapter we will attempt to navigate this illustrious terrain, with additional topics of sterilisation and selective breeding, common practices that may be rooted in traditions and evolving forms of human exceptionalism. There is not one equation or decision-making process that contains a singular right answer; one where virtue is dependent on more than simply what we want for them (or arguably us), but what our dog might want. I don't provide answers here, just perspectives that can help to guide decisions made without our dog's consent when we operate under the umbrella of "dependent agency" (Donaldson & Kymlicka, 2011), or substitutive consent.

5.2 DEPENDENCY AND SUBSTITUTIVE CONSENT

There seems to be unquestionable evidence that many nonhuman animals are able to make choices in their own interest in a variety of situations as I have discussed. It is also apparent that we can observe, through behavioural measures, their preferences, and motivations, strengthening the argument for providing them with a greater sense of autonomy and ways to exercise agency. Collectively, the evidence confirms that people should make the effort to understand their dog's preferences and either respect that dog's volition or respect the surrogate choices they make for them in the position of "in their best interest."

Philosopher Tzachi Zamir (2007) suggests that the idea of a "greater good" (i.e., bettering a life in the bigger picture) could be a justified reason for us to make decisions (even unpleasant decisions) for other animals. To highlight this point, he suggests that in the example of dairy cows, if removing the calf from their mother at an early age is more of a short-term suffering than it is a long-term harm, then separating them may be considered adequate in exchange for the chance to be alive at all. Therefore, he is saying that decisions such as these are, *allegedly*, in the animal's best interest overall. Similar arguments appear in dog training circles to defend the use of a shock collar, choke chains, or other aversive, threatening, painful devices, and coercive techniques. This would also be an argument for the "simple" use of restraint or physical force to garner compliance and control in situations to administer vaccinations or medical procedures (because it's for the "greater good."). The argument by such proponents is if the use of a bark-controlled shock collar means that the dog gets to remain part of the family and not be rehomed, then it could be considered to be in the "best interest" of that dog. However, this is a poor argument for its abuse, and such a justification certainly lacks sagacity. Not only does it neglect to consider the emotional components of why the dog is barking, it fails to consider their needs at all and simply forces a choice to appease our own senses. If we look at this same argument from a different (less human exceptionalist) perspective, it would not be a rational explanation to exploit humans in the same way. Donaldson and Kymlicka (2011, p. 93) say

> Imagine someone who justifies removing a child's vocal cords on the grounds that they would have to listen to that child's crying and screaming, and that even without vocal cords the child's life is still worth living. In the human case, we don't accept that the value of existence to the child justifies such harms.

Zamir's argument simply legitimises and institutionalises the subordinate status of nonhuman animals and doesn't consider anyone's best interest but our own. This is not to say that dogs cannot ever benefit from our gift of forethought and our ability to make decisions in their best interest—they can and do. We simply need to be mindful that these decisions are not in the guise of being in the best interest of our dogs when, in reality, they are biased by our own conveniences.

Obviously, sometimes the choices we make coincide with the dog's preferences at that moment. Those are easy choices, such as going for a walk or playing together. However, for reasons of their own safety or a better long-term outcome, a human's choices may override a dog's personal volition. For example, we know that if our dog eats chocolate or chases a cat into traffic, they are going to likely end up suffering (and even potentially causing the suffering of another), so we choose to prevent or mitigate those actions for their own safety and to thwart future suffering. If we are to make decisions on behalf of dogs, we must first and foremost ask whose best interests are considered (ours or the dog) and can we ask for their consent and respect that decision? If the answer is yes, then we should. If the answer is no, it is important that our choices should be evaluated a) in each situation (no blanket assumptions) and b) to generate an intervention that is the least invasive and least intrusive.

We can, however, use this great gift of foresight when assessing a dog's quality of life (QoL) to improve their long-term outcomes. Researchers Wojciechowska and Hewson (2005, p. 724) suggest that due to the obstacles involved in correctly evaluating the private emotions and feelings a dog may be experiencing, "objective list theory may be the most useful for QoL assessment in dogs at present"[1] in veterinary science. However, this carries its own inherent anthropocentric bias. Though the list is both objective and based on tangible items that comprise a quality life for dogs, it is still a manifestation of our own human assumptions. It is a kind of "reflective judgement," and items on a checklist are not necessarily tailored to an individual's evaluation, including the trauma some dogs may experience during various interactions due to a variety of mitigating factors. We will return to this concept in the succeeding section about end-of-life decisions.

So then how do we evaluate when the decision we make for our dog is actually better for them overall? It's a process, not a whim or assumption. Yeates (2018) suggests we might evaluate the sequence of events and conclude that the dog's desire to avoid something unpleasant in the future outweighs an immediate pleasure. These evaluations may consider whether there are additional events in the sequence that may affect the final outcome or that one might simply be more important than another. For example, my dog, Juno, doesn't like to jump in the car on her own (because she doesn't enjoy being in the car for extended periods of time). But I also think that her love of going for a hike or to the beach to meet a friend outweighs her dislike of short car trips. I base this on the behaviour I observe when she is at a favourite hiking spot and the obvious pleasure is expressing. This isn't to say that I can't actively work to improve her car experience through positive reinforcement. I can and I have. But even though she has acquired these new technical skills, and maybe she's even overcome some of the fear that she once experienced, it is still unlikely she would choose to jump into the car of her own volition. She does it because of the reward value and reinforcement history that it carries. We all have things in life we must endure or tolerate in order to get to the desirable destination, like the years and years I spent in school before being awarded my PhD. But there were still lots of extraneous positive rewards granted along the way that gave me a quality life during that period and without them, the journey to gaining a PhD would not have been worthwhile. Thus, even when our dog is performing a behaviour they may not naturally desire, we should always make it worth their time, effort, and reduce any unpleasantries to nothing more than fleeting moments that pass with hardly an eyebrow raised.

Teaching skills (operant conditioning) and the practiced process of socialisation (including habituation, desensitisation, classical conditioning, and observational learning) are necessary proceedings in order to provide a quality life for our companion dogs provided we do so with an individual's best interest in mind. It can provide clarity and information needed to live more harmoniously in a human-centric environment. One perspective is to weigh these skill-building programmes against species-specific norms. Nussbaum (2006) suggests that we analyse the decisions we make on behalf of our dogs based on species-specific (rather than individually-specific) standards. For example, if a dog is in pain, we must provide that dog with medical intervention in order for that dog to regain normal mobility because being mobile is specifically important to dogs. While this is true and important, it lacks the critical component accounting for individual variations in preference and motivation. It's simplistic and while that is helpful in some ways, it can also be problematic. In this regard, it is much like objective list theory—lacking fundamental evaluation for individuals in individual situations, the way we might do for ourselves when choosing to consent or not. And though Nussbaum's theory may fit well when considering the measure of pain, in my earlier Juno example, my decision is not based on what "all dogs like" but on what she enjoys long term. For example, not all dogs enjoy socialising with other dogs, and assuming this is true of all dogs or assuming it is true of no dogs may be harmful to an individual dog. Taking a non-social or fearful dog to a dog park could be traumatising, but preventing the social interactions of a dog who is wanting to be social may cause frustration and stress. In short, companion dogs need to be able to communicate with humans, and humans with dogs, because that is the reality of living in a multispecies society and necessary for a consensual communication to occur. And, when we are evaluating substitutive consent, we need to be able to more efficiently and effectively evaluate the individual dog in front of us to ensure that we can make the best educated decision on their behalf in situations that do not support their ability to consent.

5.3 DESEXING

To be sure, the topic of spay/neuter is a contentious one. There are those for and against spay and neuter as well as varying evidence to support the optimal age for the surgery to occur and in what specific capacity. And in all honesty, both camps make some valid points while falling short in other areas. That is because it is a complex issue, but one where human exceptionalist motives need further examination.

Spay and neuter surgeries began to become routine practice around the 1930s (Grier, 2006), and the first low-cost dedicated sterilisation clinic opened in California in 1969 with a 4-month long wait for people requesting sterilisation surgeries for their dogs. This movement was vastly spurred by criticisms about a rising number of ostensibly free-living dogs who were deemed a threat to the public (Rosenthal, 2011). Initially, the American Society for the Prevention of Cruelty to Animals was opposed to a generalised desexing programme, but this all changed in the 1970s when they began requiring spay-neuter before adoption as a way of mandating a "population control" mantra that reflected the narrative of "responsible pet ownership." Added to that is the argument that dogs live longer, healthier lives and that their behaviour (particularly males) will become more "obedient." In other words, it was considered to be convenient for us, and safer/more advantageous for them.

Traditionally, sterilisation practices have been aimed primarily at overpopulation and have presumably appeared to be successful in reducing the number of homeless dogs. Shelter intakes have indeed reduced appreciably since the 1970s, and the number of euthanasias have dropped to about 2–4 million canine and feline euthanasia annually from an estimated 20 million (Horowitz, 2019). However, to counter that point, Winograd and Winograd (2017) suggest that although it can be argued that sterilisation decreases shelter intake rates, this reduction isn't necessarily directly related to lower death rates. There are other variables to consider, such as higher adoption rates, and it is this marked change in adoptions that may be most influential over decreased death rates.

What all this suggests, then, is that routine spay/neuter programmes are not necessarily the solution to either end (particularly shelter) euthanasia, nor is it the solution to prevent rehoming or dog surrenders as has been argued extensively throughout the years. To this point, Horowitz (2019, para. 5) notes that, as of late, it is "illegal to desex a dog in Norway. Only seven percent of Swedish dogs are desexed (compared with more than 80 percent in the United States)." Additionally she notes that Switzerland's Animal Protection Act contains a clause that honours "the 'dignity of the animal,' and forbidding any pain, suffering or harm, such as would be incurred by desexing." She goes on to that that despite these marked differences, these countries don't seem to experience excessive numbers of stray dogs.

In reality, "kill shelters"—a term used to describe a shelter who will euthanise animals in their care if they are unable to find homes within a certain timeframe—are only a part of this moral dilemma of routine dog sterilisation, (though one worthy of inclusion given the number of dogs who end up in them). Winograd (2019) suggests that routine sterilisation, especially in shelters, persists largely because of the impending threat of existing "kill shelter" protocols. Could we eliminate the perceived "need" for surgical desexing by choosing to end killing in shelters? Winograd goes on to say that

> unlike humans vaccinating animals to protect them from potentially deadly diseases beyond our control, sterilisation is inflicted on animals by humans to protect them, their offspring, or other dogs from kill shelters. Adjusting our behavior, rather than manipulating their ability to reproduce, is one way to solve this ethical dilemma and it places the burden of change where it should rest: on us.
>
> (para. 12)

This circles us back to the glaring common ethical thread throughout this book: Human exceptionalism and its suitability to human lifestyles. Ultimately, sterilising dogs is convenient for us. Intact dogs can be messy or may require thought and consideration when it comes to managing their interactions. The topic of canine sex is one that's taboo, especially as people more often view them as perpetually neotenous—as their "furbabies." This is so much a problem for many people that simply seeing a dog mount and hump another dog, or anything, really, is seen by humans as obscene, despite its normalcy in dog–dog interactions including and apart from sexual encounters. How often is the topic of want or desire even broached? Probably never, despite the fact that sexual interactions for dogs, like other animals (including humans), is desired and attends to a biological need.

I think, however, it would be irresponsible to call the immediate termination of routine spay and neuter procedures without careful consideration. For one, there are those who contest the findings that spay and neuter surgeries are harmful to health, instead suggesting that these procedures decrease certain cancer risks. Others argue that dogs in fact live longer with these surgical interventions and thus, due to longevity alone, have a greater opportunity to contract diseases, skewing the results of current studies. Additionally, there are studies show that sterilised dogs suffer fewer infections such as pyometra, uterine infection, and some cancers, like mammary tumours (Urfer & Kaeberlein, 2019). One problem with drawing such conclusions is that much research on spay and neuter involves medical record reviews, which is limiting. These reviews attempt to verify the factors of various conditions (such as frequency or severity) compared with their status of spay/neuter and their age during surgical intervention. The problem with this method is that these studies show associations not causations, which tells us a little but not a lot.

Critical Corner: Some problems with medical record review-based studies, (for example, de la Riva et al., 2013; Hart et al., 2014, 2016; Sundburg et al., 2016) all reviewed medical records from a referral-based specialty hospital to examine incidences of joint problems, various cancers, and immune disorders in correlation to sterilisation status and age. All the studies reported varying degrees of risk factors associated with sterilisation and certain orthopaedic, neoplastic, and immune conditions. But does this show *causation*?

Such reviews mean there are no control variables and likely contain a biased research population (i.e., only referred patients). For example, patients managed by general veterinary practitioners may not be represented in these records at all. People who haven't sterilised their dog because they couldn't afford it may also be excluded (surely if they chose not to spay or neuter because of financial constraints, it is reasonable to assume they likely also couldn't afford to take their dog to see a specialist).

These studies have their merit, but they do not justify widespread or generic changes in spay and neuter decision-making based on health justifications alone.

There is still a lot we don't know about the effect of sterilisation on dog's physical wellness. Therefore, it's important we continue to be open to new evidence as it becomes available and open to changing our minds based on that evidence in accordance with what is best for an individual dog in terms of the best QoL we can provide for them, and it draws credence to the moral implications of the decisions we make for our dogs. At the same time, it's important to look critically at new and growing knowledge about the effects of sterilisation to determine if conclusions are valid based on what the research data tells us. This should be especially important to veterinarians and veterinary support staff, as they are often the first point of contact for dog caregivers. I'm not planning to compare the vast amount of studies that weigh the benefits and pitfalls of sterilisation on health beyond simply saying it's important to understand the newest research with a critical lens and base decisions on what is best for each individual.

Obviously knowing the health benefits plays a large part in how we evaluate the decision to sterilise our dogs or not. Or it should be. But let's shift the focus to the morality of sterilisation of dogs. What do we know? A dog cannot consent to this surgery specifically, and I think we would be hard pressed to find anyone who would say their dog would have willingly agreed to such, though perhaps if they were able to weigh the facts and figures in a logistical way, they would be in a better position to do so. Obviously, this isn't possible and frankly, that is something even we struggle with knowing, as you can see. So how do we know if we are making the right decision for our dog? Let's go back to Yeates' suggestion of an analytical evaluation of what's "best for." Does the health and welfare of that dog, over the extent of their life, outweigh the disadvantage of the temporary discomfort? How about taking away their ability to have and raise babies? How do these decisions benefit them or us?

Some argue that intent matters when making decisions about sterilisation on behalf of our dogs. What I mean is the intent may be a decision made to produce a better long-term outcome and is not necessarily malicious and that *intention* to do best is what matters most. But it's worth considering that within the ambit of human rights, intent actually does not matter at all. Let's look at forceful human sterilisation by castration, for example. Even if the intent is to reduce overpopulation and potential suffering (of the individual, of their offspring, of the planet), it would still be termed genital mutilation and illegal (and malicious) to perform without that person's consent. Some may argue that until the animal rights movement can agree to a consistent moral doctrine that all violations of the bodily autonomy should be called by the equivalent human rights terms, regardless of the *intent*, the term "rape" or "genital mutilation" or other such terms used in human rights violations should not be blithely trivialised. I disagree—such terms may in fact provide some needed leverage to remove ourselves from human exceptionalist thinking and is the exact reason why I choose the term "consent" over other synonymous or closely related terminology. If we can't stop compartmentalising our actions toward other animals and aiming to justify their oppression, we cannot move forward into an inclusive coexistence. Coincidentally, it wasn't that long ago that White colonialised families thought that slavery of Black persons was not only just, but that they were doing them some sort of favour. Maybe one day our future generations will look back at the way that we institutionalised the "othering" of nonhuman animals with shame and remorse, too.

Another argument made by some is that because one theory suggests that dogs may have domesticated themselves and thus are bound by a social contract with humans, this justifies their forced sterilisation (for example, Budiansky, 2016). That may seem far-fetched, but the argument maintains that dogs seemingly have partnered with humans for food, shelter, medicine, affection, and love, and people can insist they not be burdened with offspring they cannot take care of. What a weak (and human-centric) argument. In what capacity does that account for the individual dog in your home? (Which is who spay/neuter affects also, not just the population of dogs and human society). And as such, our dogs did not choose us—they were taken from their mother and siblings at a very young age, likely never met their father, and were placed into a home without any choice to leave. They didn't choose to partner with one particular family, that family chose to "partner" with them. Like so many arguments that aim to justify our treatment of dogs, this highly anthropocentric argument operates on the assumption that preventing dogs from reproducing is reasonable because dogs, like humans, are bound by human social contracts, which is a ridiculous and unfair.

If we look at the equation thus far, the main considerations when making decisions about sterilisation should be

1. Health implications for that individual.
2. Chances of pregnancy and how might that impact our dog, her health, and the emotional toll of having her babies removed from her.
3. Invasiveness of the procedure itself. Removal of testes and a full hysterectomy doesn't come without risk. Surgical sterilisation itself can cause infection, it inflicts pain, it can cause physical and emotional stress on the body, and it can increase the risk of diseases as mentioned earlier.
4. Physical and emotional risk of hormonal changes that also may lead to behavioural or emotional changes.

Most, if not all, of the decisions made on behalf of dogs are paternalistic; dogs are powerless to exercise true agency particularly when it comes to reproduction and sexual encounters. Thus, we are morally obligated to act in ways that suit their best interest by using the least invasive modality to help them to thrive. This may mean practices such as vasectomies, tubal ligations, or hormone injections in lieu of hysterectomies and castration. Perhaps this less invasive alterative would additionally address the concerns about the increased risk of cancer and other diseases.

I think the largest take-away from the discussion on desexing is that there is prejudicial resistance to discourse and certainly action concerning the morality of spay/neuter. As such, it is time we have a conversation that considers the original antiquated justifications and establishment dogma for sterilisation of dogs and weigh them against the contemporary truths. We should examine the advantages and disadvantages of spay and neuter surgeries against what it means to allow dogs to remain sexually whole. This conversation is required if we are ever to come closer to ensuring that the choices we make on behalf of dogs—those decisions made without their consent—are at the very least the very best choices we can make based on current evidence.

5.4   SELECTIVE BREEDING: COMMODIFICATION OF DOGS

This topic is perhaps one of the most contentious I will broach, and one of great importance. First off, well selected (healthy and resilient) parents usually make for well-adjusted offspring, who in turn become companion dogs best suited to fit into the lifestyles of human homes. The better the fit, the higher the chances that dog will not end up being rehomed or relinquished (or euthanised). This is because they are selected to be social and emotionally resilient and physically healthy (recall in Chapter 4 the link between physical health and behaviour). That means less rehoming, less behavioural euthanasia, and it makes sense, right? Yes. But like all things considered in a paternalistic relationship that is rooted in human exceptionalism, it's complex.

Critical Corner: Before we explore the topics related to selective breeding, artificial insemination, or forced pregnancy, let me ask you this: Would you consider the things we do to breed dogs to be "rape"? Why or why not? We'll come back to this question because it's both very confronting and even potentially triggering but worthy of discussion.

To be forthright, this section is a discussion of the morality and ethics of selective breeding, not necessarily a discussion of puppy or parent health and fitness. Though these are important topics, this book will focus on the discourse around the doctrines and morality of the *practice* of selective breeding itself and whether or how it is positioned in discussions of consent.

Obviously, some dogs have sex and become pregnant without human intermediation. Juno was a product of such a situation. Her mother set out on a rural escapade, crossing paths with the neighbour's elderly pointer cross. Approximately 2 months later, Juno and her brothers were born. But this is rare in most Western cultures and often scorned for its irresponsible nature, returning full circle to the previous discussion on sterilisation and management and the narrative around ir/responsible ownership. Situated at another extreme are puppy mills, where females are forced to reproduce in ways that are all too familiar in other practices of animal commodification such as the factory farming of dairy cows and pigs who are forced to have countless babies as if their bodies are mere material objects. Somewhere near the middle sits the dog breeder—those people who select a female and a male for their heritable traits (usually health, aesthetic, breed, and temperament) to reproduce those same purpose-bred characteristics in their offspring. Those offspring may already have certain futures carved out for them based on their breed, breed type, or lineage, including sport, human service roles, or companion.

The commodification of dogs is embedded within the discourse of dog breeding infrastructures. The simple action of "producing" dogs solely to profit from their existence metamorphoses dogs from companions to commodities, delineating their unique position within human families and their utilitarian value to humans who own them. According to Hens (2009), humans have special obligations to dogs that are shaped by their status as family members, including caring for their physical and emotional well-being. My earlier research shows that people consider their dogs to be family, though not necessarily equal to human family (Jones, 2022). Obviously, there is a reasonable measure of cognitive dissonance present if we can view dogs as companions and family (even given a lower value or worth) while concurrently advocating for their commodification (breeding, sale, ownership). Pragmatically, breeding dogs clearly differs in some ways from other animals farmed specifically for humans to benefit from their bodies and/or what their bodies produce, but there are also parallels that exist between those domains, and such commodifications are all affected indignantly by the human exceptionalist paradigm.

Anyone who has lived with dogs (and even those who don't) could never dispute dogs as sentient beings—clearly, they are. However, people who buy or breed dogs appear to struggle to attribute any extrinsic value to them while simultaneously viewing dogs as having intrinsic value (Croney, 2019). Croney (2019, p. 233) says that it is likely people who buy and breed dogs "hold beliefs about animal–human relationships that facilitate contradictory or compartmentalized thought processes about dogs." They go on to write that for some people, their views "may allow them to conclude that producing dogs (even under sub-par conditions) brings about better societal consequences than not breeding them at all." This reasoning would be consistent with consequentialism—the theory that whether an action is deemed to be justifiably "good" is contingent on whether it has a favourable outcome. But good for who? Certainly this favours the human side of the equation.

It could be argued that outside of ART origins, if domestication is a sort of captive–master relationship that is preserved by selective breeding, then buying and selling dogs is not inherently wrong if they have been bred and cared for in ways that makes a genuine effort to protect their welfare. That rather naïve and supercilious claim is, suggests that dogs are created by us, for us, and therefore, we can do what we like to them. This is especially true for large, commercial breeding operations where it is tough to discern any argument that supports the idea that having to live solely to produce puppies for someone else's profit is in any dog's interests. And if we argue for "ethical breeding," including breeding for health, considering both the parents' and offspring's welfare, where do we draw the moral line of exploitation? Is artificial insemination better or worse than live breeding? Is either choice ethical if breeding is a human choice for a human outcome?

Artificial insemination (AI) is a common practice with breeding dogs. To be noted, there are two kinds of AI, intrauterine (requiring surgery) and intravaginal. Surgical AI in dogs—insemination under general anaesthesia by way of a laparotomy, adding thawed semen into the uterus through exteriorisations—has been banned in the United Kingdom since 2019 and considered an unethical practice in many other countries. Intravaginal AI requires inserting thawed semen into the vulva by way of syringe and catheter, which are both wholly invasive for the female, not to mention the male whose semen is being collected and implanted. A colleague described to me a while back a situation where her intact male dog was made to become sexually aroused before the vet manipulated his genitals so he could contribute sperm to an unknown female halfway around the world. And that female likely underwent some form of vaginal penetration and unknowingly became impregnated, gave birth, and then had her babies taken from her at 8–10 weeks of age.

There are valid points on both sides of this argument depending on the perspective you decide to take; the topic is multifaceted, and we make a lot of decisions on behalf of dogs which fall both within and outside the realm of "best interest." By Yeates' measure of "best interest" explored at the onset of this chapter, could breeding dogs ever be considered a "best interest" outcome? Whose interest does it serve? Let's consider:

1. Dogs do not get to have sex to satisfy any sexual needs or desires (generally).
2. They may become impregnated through AI practices (or have their semen harvested) without consent or having any bodily autonomy.
3. They are forced to carry babies for approximately 2 months, and surely, we have no way of knowing if this is something that any mother would want.
4. They are then required to share raising responsibilities with humans, and their babies are removed from them.

One could surmise that these practices could have residual traumatic effects, and I feel confident in saying that we, at the very least, are not well equipped to decide whether any individual would in fact want or choose to go through any or all of those steps listed.

> **Critical Corner:** In what way is selective breeding considered the best interest of dogs? Does it at all? What are the benefits and drawbacks of selective breeding when we consider the potential good lives of those being born?

No discussion about the ethics of selective breeding would be complete without discussing how aesthetic trends affect even the most ethical practices of selective breeding. Let's look at the hot topic of brachiocephalic dogs, for example. They are bred extensively for consumer demand—people find flat, baby-like faces appealing. But their skull shapes pose a surfeit of problems, like brains being too big for the space their skull provides and severe breathing issues. One study (of several) from the United Kingdom found the popular breed of French bulldogs have a greater probability of physical health problems like ear infections, dermatitis, and difficulty birthing naturally (O'Neill et al., 2021). And as certain dogs are bred in correspondence to changing trends, the ethics of such practices are being more closely examined. The Australian Veterinary Association has explicitly advised against breeding brachycephalic dogs and any breeds with a higher risk of spinal abnormalities—among other physical health-affecting complications—while other countries have started to advocate for more regulations in order to reducing the risk of suffering. For example, animal rights advocates in Norway argued against continued practices of inbreeding, leading to the exaggerated facial structures of bulldogs, which led to a breeding ban (that has since been appealed).

"Responsible" breeding is certainly favoured over haphazard breeding practices such as those labelled "backyard breeders," though there are varying shades of ethical complications when delineating exactly what a doctrine of responsible breeding may represent. Selective breeding begets ethical consequences that spur difficult questions that come with difficult answers. However, breeding remains a problematic practice regardless of welfare standards and responsible practices if we are ever to consider whether dogs should have the right to consent in situations that may violate their bodily autonomy.

5.5 INVASIVE MEDICAL PROCEDURES

Juno has not had an easy go of life, medically speaking. In 2022 she became very ill. A fever of unknown causes spiked overnight, her bloodwork showed inflammation of an unknown origin, and it appeared her abdomen was painful to the point of persistent trembling and some ataxia, but we couldn't pinpoint the problem. We chose to do an ultrasound and found she had an enlarged spleen, but because the fine-needle aspirate (FNA) was unsuccessful and after no resolve to her symptoms, we chose to do a splenectomy and exploratory laparotomy. She was clearly suffering and had lost a significant amount of weight in a short period of time. Though the biopsy was inconclusive, she has not had a relapse and has been healthy and happy. Fast forward to 2023, Juno presented with some odd behavioural changes that were not related to environmental factors (as in, they were not triggered by environmental stimuli). These episodes were very unsettling for her, and she would desperately seek my connection and support post "episode." These episodes were presumed neurological, though not presenting as a typical grand mal seizure but rather as abnormal phantom fly biting followed by growling, lip smacking, dilated pupils, and clear appeasement signals, that was without intent or conscious control. We began with the least invasive testing first, moved forward to an ultrasound, then a CT scan, an MRI, and eventually a spinal tap and a bile acid test. We ruled out every possible thing that the tests suggested "could" be causing her episodes but ultimately landed on idiopathic seizure disorder. Unbeknownst to me, partial focal seizures affect a small localised part of the brain and can manifest in various behaviour or movements with the patient having only partial awareness but no control. During the entire process, I certainly felt guilt and apprehension, as each test required full sedation, and I didn't have the flexibility to ask for her consent to inject the sedative (cooperative care was not part of this process for the very reason that "no" was simply not an option she was provided). Thus, I didn't ask for consent for any of it: Not whether we could sedate her, not whether we could do the tests, nor whether we could treat the cause of her illness. Maybe it's what I wanted for her, or maybe it's what I would want (as a human, of course) in a similar situation. I did it because if there was a treatment that may well grant her a fuller life that was no longer burdensome, painful, or causing her to feel unwell, of course I would choose that for her. And I would aim to get there in the way that caused her the least amount of stress: *asleep*. In the end, we did manage to rule out any major and more sinister causes of her symptoms, such as a portosystemic shunt, a brain tumour, or meningitis. And I certainly don't regret the decisions I made on Juno's behalf during either major medical event. My aim was to be able to accurately assess her subjective well-being, minimise her stress during diagnosis, and make her life better, longer, and happier.

You may recall, in Donaldson's and Kymlicka's political animal rights book, *Zoopolis* (2011, p. 105), they express that it is through the bonding process with our dogs and through our responses to their expression of needs, wants, and desires that we can foster a "dependent agency" where dogs are able to form relationships with humans that help them to "manifest a subjective good, to cooperate" and to be "participating agents." It is always a measure of their subjective experience that should be considered when assessing consensual interactions (Yeates, 2018). This theory shares similar qualities of substitutive consent (see Chapter 1), in which a legally authorised person makes decisions on behalf of someone with the intent to do so in that person's best interest. Hence, when navigating dependent agency, we should facilitate experiences that lead to desirable outcomes and avoid undesirable ones. Yeates (2018) says, "each scenario, or type of scenario, needs to be evaluated in their particulars such as species, situation, content and individual" (p. 182). Though Juno didn't consent (or not consent) in both major medical events I just described, I made cues clear through the process of earlier teaching/learning strategies and based on previously constructed and foundationally sound communication systems between me and Juno, that the sequential process of handling, at the very least, would have been clearly demonstrated. Though I didn't ask for her consent (via the use of a consent-behaviour—her chin rest) as outlined in Chapter 4, I was still able to practice handling in advance of such emergent situations and thus reduce her stress through predictability and emotional support. So while I didn't ask her to consent for her sedation injection, she knew, through previous practical experiences, what was about to take place. For example, using consent-based interactions, I practiced a light restraint paired with the predictive cue "hug" and the poke of an intramuscular needle (using a blunt needle) paired with the cue "poke." Using these specific cues, I was still able to provide her with valuable information about what was about to take place during these instances where consent wasn't an option. Fear of the unknown is a powerful aversive and will almost always cause feelings of anxiety or stress. Eliminating this fear by providing predictability was essential. And though her consent wasn't given, this situation was the exception, and hopefully there are few situations where this is required.

As a result, Donaldson and Kymlicka (2011) concede that the concept of dependent agency is a trust-based dependency and is a way for agency to be exercised through someone who has the skills and knowledge to act on another's behalf in their best interests. So although we are likely unable to ask for consent in life-altering medical emergencies we can still offer information based on past learning, and often dogs will assent to handing with minimal physical intrusion.

Figure 5.1 **Juno returning home from her splenectomy, 2022, with an intravenous catheter still inserted.**
Source: **Photo credit: Erin Jones.**

5.6   END-OF-LIFE DECISIONS

Back in 2016, my husband and I, and our 16 year-old dog, Monday, who had been suffering from cognitive dysfunction for nearly 2 years, moved in with my mom who was also suffering from later stages of early onset dementia. She specifically had a form known as semantic dementia and was unable to fully care for her home or self without additional assistance. Within that same year, Monday's brain functioning began to show a noticeable decline, and my mother's executive functioning had virtually faded into an elusive black hole of some kind of cerebral abyss. She didn't recognise most items or foods and would only eat the same meals at the same time every day. She was unable to show emotion though early on she showed depression, and I imagine it was still there, and she just couldn't express the way she was feeling.

During that time, I created a list for Monday. I ranked her favourite pleasures in life: Playing fetch, swimming, long walks, food, attention, car rides, and the park. She progressively became nervous and confused and eventually would forget where she was going or how to play fetch. She would stop on walks and not know what to do next, and it was distressing for her. Evenings were difficult too, and her sleep schedule changed. She became incontinent and was restless most of the night. In vain, I also made a list for my mom. It was worthless in the larger scheme of life and death; it had no resolve beyond long-term respite care. When Monday became overwhelmingly unhappy, stressed, anxious, and unable or unwilling to do any of her favourite activities, I had the distinct privilege to end her discernible mental agony. I didn't have that same privilege for my mom; I simply had to do everything I could to prolong her life no matter how much she suffered in her meagre existence and how miserable she was feeling inside. She ended up passing away from congestive heart failure, and I don't know how I feel about that conclusion. She was unable to even communicate how she was feeling leading up to the event, and as such, I wasn't even able to comfort her until it was too late. But for Monday I could. Sure, we went back and forth over making end-of-life decisions until the day she no longer wanted to eat, and just like that, we came to the last check on the list, and we felt it was time to say goodbye. I felt honoured I was able to assist her through this transition and fully be there with her during it. Both deaths were difficult for me, but one was, I think, the compassionate option (although I am sure some people may disagree). Assisting dogs in death, however, is complicated, and the reasons are varied.

Once regarded as "disenfranchised grief"—generally not supported by society as a genuine and laudable reason to grieve—the death of a valued dog companion is now more commonly recognised as being as significant as losing a human friend or family member. Palliative care for dogs has developed as a progressing field in veterinary medicine. However, when weighed against the rights and laws supporting end-of-life decisions for humans (as dubious as they may or may not be) end-of-life decisions for dogs in our care do seem to be more fickle, though sometimes more malleable in

Figure 5.2 "Grandma" Monday enjoying some time at the playground.
Source: Photo credit: Erin Jones.

both positive and negative ways. Veterinarian professionals often describe euthanasia as representing "the best and the worst" of the profession (Hartnack et al., 2016). Nonhuman animal death, particularly euthanasia, can be morally complex and can be immensely stressful, affecting the mental wellness of those involved; thus it poses enormous challenges for everyone (Yeates & Main, 2011).

Veterinary professionals are often forced to compartmentalise their own feelings and counter the well-being of a dog who is chronically ill, struggling behaviourally, or simply unwanted with the needs of their human caregivers. At times, the dog may have a completely treatable condition though their caregiver is unable to financially support immediate or ongoing treatment, in which case the veterinarian is forced into performing an "economic euthanasia" (Boller et al., 2020). For many people, this can become a sizeable obstacle when making medical decisions on their dogs' behalf. In the example of undesirable behaviours, some caregivers may be disinclined or incapable of investing the time and money it would require to appropriately address these behaviour "problems" with a professional, or over-taxed shelters may not have the capacity to intake or help to rehome unwanted dogs, so euthanasia may be the outcome either way.

The reasons people may choose to assist their dog in death is varied. In 1996, data was collected from across the prairie provinces in Canada showed that "old age" was listed as the top reason for euthanasia, followed closely with terminal sickness at 30.2% (Gorodetsky, 1997). However, "aggressive behaviour" made up 9% of euthanasias, and other "behavioural abnormalities" made up 4.2% (Gorodetsky, 1997). In a similar study of dogs in the United Kingdom, the most frequent reasons for euthanasia overall were found to be for medical diseases and disorders. However, when that data is further reduced by age, the most frequent cause of euthanasia for dogs under 3 years of age included "behavioural abnormalities" (O'Neill et al., 2013), which may or may not have an underlying physical differential diagnosis to be considered. There is sizeable evidence to show that "behavioural problems" are considered a principal reason for the euthanasia of dogs (O'Neill et al., 2013; Pegram et al., 2021; Fatjó et al., 2006; Boyd et al., 2018) with several studies reporting that "behavioural problems" comprise between 2–39% of canine euthanasias (Michell, 1999; McMullen et al., 2001) and up to 50–70% within animal shelters (Fatjó et al., 2006; Overall, 2013).

It is well known that both biogenetic and environmental influences shape the behavioural development of dogs, and there is nascent evidence that supports proper socialisation and habituation practices to decrease the occurrence of "problem behaviour" in adult dogs (Westgarth et al., 2012). Not only can proper skill building and learning techniques help dogs to become well-adjusted members of society (even those who are predisposed based on their genetics), social skills can reduce the incidence of a dog developing behaviours that may become detrimental to their well-being, either directly or indirectly. For example, if behaviours are associated with negative emotional states and/or aversive responses from humans, it can dramatically affect their quality of well-being through chronic stress and/or stressful and unfavourable consequences. It may also thwart opportunities to learn resilience and hinder the ability to thrive in their environment. Teaching social skills, coping, and resilience skills, while providing opportunities for puppies to develop their brain function through positive experiences and outcomes may ultimately reduce the occurrence of "behavioural euthanasia." Moreover, supposing human exceptionalism influences the behavioural expectations we have of dogs would suggest that some behaviours may be characterised as "undesirable" when they are in fact quite normal species-specific behaviour (Jones, 2022). Even behaviours labelled as "aggressive" are often within the normal realm of canine communication. Therefore, it's quite often the caregiver's perception of desirability or their belief of what demarcates "good" or "bad" behaviour that may affect the likelihood for behavioural euthanasia. And these decisions are based on human convenience rather than based on an individual dog's "best interest." Perhaps education in dog behaviour will work to prepare people properly before acquiring a dog or dissuade them from adding a dog to their family when they are not willing or ready to compromise on their lifestyle.

When it comes to highly painful or terminal illness and life-altering disabilities that will affect the quality of that dog's life in significant ways, the decision might seem less burdensome, though it can be hard to know the "right" time. Someone once told me, "it's better one week too early than one week too late." I'm not sure if that would always be the case or how we can judge time on either side with any tangible precision, but the sentiment is that we can end that life before they suffer for it. And after the situation with my own mother and with Monday, I would agree. But it's not an easy decision, with humans or other animals in our care. It's both ambiguous and shrouded with much ethical debate.

"One week too soon," to be clear, is about that dog's reality, not one based on our own personal burdens of caring for that individual. On the other hand, there remains an argument that euthanasia is a harmless practice because dogs can't anticipate death or perceive a future. However, we may not know, or ever know, exactly what the dog experience is: How they may grieve, their vision of the future, or their awareness of death. But as Pierce suggests in her book, *Run, Spot, Run* (2016), dogs are likely not devoid of any of these concepts. And just because we haven't (yet) been clever enough to unearth such evidence, for or against, is not a reason to assume that the ability doesn't exist at all. And although there really isn't conclusive evidence about the complexity of a dog's aptitude to reflect how their future might unfold, Horowitz (2010) points out that there are indeed studies that show other animals have an "autobiographical consciousness." For example, in university I learned that the Western scrub jay caches food in hundreds of places for later use and remembers exactly what, how much, and where they have cached these items. I'm not sure this is an important or worthwhile practice for dogs, but I can testify that Juno can easily anticipate near future events. She knows what time I normally go to bed, and if I deviate, she lets me know! She also knows about what time my husband, Mike, should be getting home from work and begins to anticipate his arrival by preparing which toy she will greet him with upon his arrival.

Perhaps looking to the arguments about human euthanasia will shed some light on how we may want to examine the virtues and iniquity of dog euthanasia and how this may correspond with an application of canine consent. Within the discourse of human euthanasia, there are various "types" of euthanasia drawn. These include "voluntary euthanasia," "assisted suicide," and "medical assisted suicide," all which require, by law, a person's fully contracted consent. The conditions in most countries prohibit assisted euthanasia, although there are a small number of countries, states, provinces, or regions where medically assisted euthanasia is lawfully sanctioned under very strict and specific conditions. Those who advocate for euthanasia generally contend that it is a violation of human rights to disallow people to choose their own death and that it is inhumane to unnecessarily allow for continued pain and suffering—thus ascribing the expression "mercy killing." The question, then, is whether euthanasia is ethically worse than the withdrawal or withholding of medical treatment to allow for a natural death (anecdotally I would argue this transpires commonly in nonhuman animal care as well). Opponents of euthanasia contest these views, suggesting that euthanasia is equivalent to murder and likewise, exploits a person's autonomy and rights.

Keep in mind, the current climate of our human exceptionalist society means we live in a world where the killing of other animals is not just normalised, we even breed and raise many animals for killing purposes. Consent is not even a word that would be considered seriously in such conditions (though nor would *murder*). Conversely, non-voluntary euthanasia in human medicine is performed when the patient is incapable of consenting, and this, lawfully, will always amount to *murder*. Does one life, based solely on species, deserve more consideration than another?

> **Critical Corner:** Dogs sit in a liminal position: Not reproduced for consumption (in Western society), but not lawfully equal to human rights/protection of life. Is choosing to euthanise a dog based on a non-medical condition such as their behaviour (always involuntary) considered murder? Why or why not?

There are countless opinions that support and dissent human euthanasia. Let's examine these in comparison to animal rights (which, by the way, are virtually non-existent) but may help us to make distinctions important for the discussion.

1. First, those who support human euthanasia argue that a person has the right to decide how and when they ought to die, fundamentally based on the tenets of being autonomous (Bartels & Otlowsk, 2010). Do dogs have this same right? No. Should they? I don't think that a dog living in a human home will ever live autonomously or be able to make autonomous decisions within medical decisions concerning their physical health and well-being. When and should we be making these decisions on their behalf? Humans are able to use information that is inaccessible to dogs in order to make calculated decisions, though there is certainly no regulation on how these decisions are made. I would always choose to treat or end suffering over allowing my dog to feel unwell or chronically and uncontrollably painful, but that is not the case in every situation and within every human–dog relationship.

2. Secondly, it is argued that the benefits of relieving the pain and suffering a person may be experiencing through performing euthanasia outweigh any harm (Norval & Gwyther, 2003). Supporters of medical assisted euthanasia believe that fundamentally, the moral virtues of compassion and clemency are enough justification to facilitate euthanasia (Norval & Gwyther, 2003). This wholly applies to our dogs as well, but the problem lies less in whether it is in fact more compassionate, but instead on deciding what information we use to know that it's more compassionate. And if it truly is more compassionate, which in some cases it likely is, then why is this not the case for humans too? We are all animals who feel pain, and we all have the ability to suffer both physically and emotionally.

3. Thirdly, advocates of human euthanasia argue that "active" euthanasia is not more ethically problematic than "passive" euthanasia—the withdrawal or withholding of medicinal interventions resulting in that person's death. It would be difficult to argue that allowing a person to die a slow and painful death or live for days in an induced coma is a better outcome than ending that suffering. James Rachels (1997) says that medical assisted euthanasia is more humane than passively withholding medical treatment as it is "a quick and painless" procedure, whereas the latter may foment "a relatively slow and painful death."

However, adversaries of medically assisted euthanasia contend that there is a distinct ethically drawn line between intentionally ending a person's life and withdrawing or withholding treatment with the purpose of letting them die naturally. If we choose to let a person die from their terminal illness, it removes the moral culpability by allowing the disease to be the reason for death (Kerridge et al., 2009). Is this a burden for the patient or a burden for the decision maker?

I am certainly not saying it is easy to make the decision to end the life of someone we love, but to choose not to because that decision is too emotionally painful seems rather selfish. Notable to the argument against medically assisted euthanasia is society's view of the sacrosanctity of natural (human) life based in secular and/or a religious origin (Walsh et al., 2009)—that (human) life must be respected and preserved (Bartels & Otlowski, 2010). This does not hold true for society's current view for other animals, thus there is little discussion about preserving their sanctity of life. Such an argument would obviously hamper our desire to use other animals for our own conveniences and situate other animals in a position worthy of equal consideration rather than placed clearly in a position of subservience.

5.7 CONCLUSION

The topic of consent in end-of-life decisions, medical decisions, and other life-altering (and potentially life-betterment) decisions may rest within two distinct and viable theories: First, Donaldson's and Kymlicka's (2011) theory of "dependent agency" (mentioned earlier throughout this book) and second, with the concept of substitutive consent. Yeates suggests that we use the understanding gleaned through the medical and scientific evidence to guide us through the calculations needed to assess both quality of life and outcome, and make decisions that are in the best interest of the individual in our care when consent is not an option.

The problem lies in making the decision to euthanise based on financial or lifestyle situations as opposed to the best outcome for that individual. It also lies within the ethics of providing that individual the right to consent when it's appropriate to do so and to teach them the skills they need to live a successful and happy coexistence. This brings us full circle to the discussion on sterilisation and selective breeding and how it affects the individual and the bigger picture for future individuals. But in any and all decisions we make on behalf of the dogs we love, it may be worth considering how we might be able to impact their lives in positive ways—ways that dogs who live in homes can potentially benefit over those who don't. This is not based on the question of consent or autonomy but on how to make these decisions without human bias.

NOTE

1. Objective list theories are theories of well-being based on a list of objective properties that ultimately benefit the individual's overall life. Other theories of well-being include hedonism (an overall balance of pleasure over pain) and desire satisfaction theories (a fulfillment of an individual's desires). However, objective list theories instead lay claim that there are things that improve the well-being of someone's life regardless of whether they are pleasurable or desired by them.

REFERENCES

Bartels, L., & Otlowski, M. (2010). A right to die? Euthanasia and the law in Australia. *Journal of Law Medicine*, *17*(4), 532–555.

Boyd, C., Jarvis, S., McGreevy, P. D., Heath, S., Church, D. B., Brodbelt, D. C., & O'Neill, D. G. (2018). Mortality resulting from undesirable behaviours in dogs aged under three years attending primary-care veterinary practices in England. *Animal Welfare*, *27*(3), 251–262.

Budiansky, S. (2016). *The truth about dogs: The ancestry, social conventions, mental habits and moral fibre of canis familiaris*. Hachette UK.

Croney, C. C. (2019). Turning up the volume on man's best friend: Ethical issues associated with commercial dog breeding. *Journal of Applied Animal Ethics Research*, *1*(2), 230–252.

de la Riva, G. T., Hart, B. L., Farver, T. B., Oberbauer, A. M., Messam, L. L. M., Willits, N., & Hart, L. A. (2013). Neutering dogs: Effects on joint disorders and cancers in golden retrievers. *PLoS One*, *8*(2), e55937.

Donaldson, S., & Kymlicka, W. (2011). *Zoopolis: A political theory of animal rights*. Oxford University Press.

Fatjó, J., Ruiz-De-La-Torre, J. L., & Manteca, X. (2006). The epidemiology of behavioural problems in dogs and cats: A survey of veterinary practitioners. *Animal Welfare*, *15*(2), 179–185.

Francione, G. L. (2007). Reflections on "animals, property, and the law" and "rain without thunder". *Law and Contemporary Problems*, 70(1), 9–57.

Gorodetsky, E. (1997). Epidemiology of dog and cat euthanasia across Canadian prairie provinces. *The Canadian Veterinary Journal*, 38(10), 649.

Grier, K. C. (2006). *Pets in America: A history*. University of North Carolina Press.

Hart, B. L., Hart, L. A., Thigpen, A. P., & Willits, N. H. (2014). Long-term health effects of neutering dogs: Comparison of labrador retrievers with golden retrievers. *PLoS One*, 9(7), e102241.

Hart, B. L., Hart, L. A., Thigpen, A. P., & Willits, N. H. (2016). Neutering of German shepherd dogs: Associated joint disorders, cancers and urinary incontinence. *Veterinary Medicine and Science*, 2(3), 191–199.

Hartnack, S., Springer, S., Pittavino, M., & Grimm, H. (2016). Attitudes of Austrian veterinarians towards euthanasia in small animal practice: Impacts of age and gender on views on euthanasia. *BMC Veterinary Research*, 12, 26.

Hens, K. (2009). Ethical responsibilities towards dogs: An inquiry into the dog–human relationship. *Journal of Agricultural and Environmental Ethics*, 22, 3–14.

Horowitz, A. (2010). *Inside of a dog: What dogs see, smell, and know*. Simon and Schuster.

Horowitz, A. (2019). Dogs are not here for our convenience. *The New York Times*. www.nytimes.com/2019/09/03/opinion/dogs-spaying-neutering.html

Jones, E. E. A. (2022). *Silent conversations: The influence of human exceptionalism, dominance and power on behavioural expectations and canine consent in the dog-human relationship* [A thesis submitted in partial fulfilment of the requirements for the degree of Doctor of Philosophy in Human-Animal Studies, University of Canterbury]. https://libcat.canterbury.ac.nz/Record/3183157

Kerridge, I., Lowe, M., & Stewart, C. (2009). *Ethics and law for the health professions* (3rd ed.). Federation Press.

McMullen, S. L., Clark, W. T., & Robertson, I. D. (2001). Reasons for the euthanasia of dogs and cats in veterinary practices. *Australian Veterinary Practitioner*, 31(2), 80–84.

Michell, A. R. (1999). Longevit of British breeds of dog and its relationships with-sex, size, cardiovascular variables and disease. *Veterinary Record*, 145(22), 625–629.

Norval, D., & Gwyther, E. (2003). Ethical decisions in end of-life care. *Continuing Medical Education*, 21(5).

Nussbaum, M. (2006). *Frontiers of justice: Disability, nationality, species membership*. Harvard University Press.

O'Neill, D. G., Church, D. B., McGreevy, P. D., Thomson, P. C., & Brodbelt, D. C. (2013). Longevity and mortality of owned dogs in England. *The Veterinary Journal*, 198(3), 638–643.

O'Neill, D. G., Packer, R., Francis, P., Church, D. B., Brodbelt, D. C., & Pegram, C. (2021). French bulldogs differ to other dogs in the UK in propensity for many common disorders: A VetCompass study. *Canine Medicine and Genetics*, 8(1), 1–14.

Overall, K. (2013). *Manual of clinical behavioral medicine for dogs and cats–e-book*. Elsevier Health Sciences.

Pegram, C., Gray, C., Packer, R., Richards, Y., Church, D. B., Brodbelt, D. C., & O'Neill, D. G. (2021). Proportion and risk factors for death by euthanasia in dogs in the UK. *Scientific Reports*, 11(1), 1–12.

Rachels, J. (1997). Active and passive euthanasia. In N. Jecker, A. Jonsen, & R. Pearlman (Eds.), *Bioethics: An introduction to the history, methods, and practice* (pp. 77–82). Jones and Bartlett Publishers.

Rosenthal, C. M. (2011, July 6). When did U.S. get first spay/neuter clinic? *My San Antonio.* https://blog.mysanantonio.com/animals/2011/07/when-did-u-s-get-first-spayneuter-clinic/

Sundburg, C. R., Belanger, J. M., Bannasch, D. L., Famula, T. R., & Oberbauer, A. M. (2016). Gonadectomy effects on the risk of immune disorders in the dog: A retrospective study. *BMC Veterinary Research*, *12*(1), 1–10.

Urfer, S. R., & Kaeberlein, M. (2019). Desexing dogs: A review of the current literature. *Animals*, *9*(12), 1086.

Walsh, D., Caraceni, A. T., Fainsinger, R., Foley, K., Glare, P., & Goh, C. (2009). Euthanasia and physician-assisted suicide. In Saunders (Ed.), *Palliative medicine* (1st ed., pp. 110–150). Elsevier Health Sciences.

Westgarth, C., Reevell, K., & Barclay, R. (2012). Association between prospective owner viewing of the parents of a puppy and later referral for behavioural problems. *Veterinary Record*, *170*(20), 517.

Winograd, N. (2019, September 4). Debating the morality of spay/neuter. *Nathanwinograd. com.* Retrieved April 21, 2023, from www.nathanwinograd.com/debating-the-morality-of-spay-neuter/

Winograd, N., & Winograd, J. (2017). *Welcome home: An animal rights perspective on living with cats and dogs.* Almaden.

Wojciechowska, J. I., & Hewson, C. J. (2005). Quality-of-life assessment in pet dogs. *Journal of the American Veterinary Medical Association*, *226*(5), 722–728.

Yeates, J. W. (2018). Why keep a dog and bark yourself? Making choices for non-human animals. *Journal of Applied Philosophy*, *35*(1), 168–185.

Yeates, J. W., & Main, D. C. J. (2011). Veterinary opinions on refusing euthanasia: Justifications and philosophical frameworks. *Veterinary Record*, *168*(10), 263.

Zamir, T. (2007). *Ethics and the beast: A specialist argument for animal liberation.* Princeton University Press.

CHAPTER 6

CONSIDERING DOGS

A DOG-INDEXED DEFINITION OF CONSENT

6.1 INTRODUCTION

And so the story goes, back in 1923, a family from Oregon were visiting relatives in Indiana, USA, when their 2-year-old dog, Bobbie, bolted after being confronted by three other dogs. After an extensive search and fearing the worst, the distraught family had to return to Oregon (over 4,000kms away), without Bobbie. Six months later, after a cold, harsh winter, Bobbie turned up at his home in Oregon, filthy and emaciated. For this, he is affectionately known as "Bobbie the Wonder Dog" (John, 2011). This isn't the only story of its kind. The internet is full of stories of "loyalty" and love, dogs who rescued babies from drowning or found help for their injured or stranded caregivers. Many of these real-life tales mirror those of the protagonist Lassie (or if you're Canadian and as old as me, those akin to the type of rescue fables of archetype "The Littlest Hobo"). There are of course many viable and probable answers for why a dog may behave in a particular way. The behaviour analyst in me looks at the likely motivating operations and conditions of their environment, paying special concern to behaviour's function and form. But whether these actions are simply a complex structure of stimulus responses or based on loyalty and affection or something else, I think they do show some degree of mutual love between humans and their dogs. I don't think there is anyone from any branch science of animal behaviour science that could argue that the dog–human relationship doesn't matter (no matter how that relationship is defined or how love is defined, for that matter!) The emotional lives of dogs are irrefutable.

To that point, dogs are an important part of many human lives, and I hope to see a society that moves toward being more inclusive of dogs, tolerant of their differences, educated about their language, and respectful of their privacy and bodily autonomy, especially in public. However, the shroud of human exceptionalism is embedded deeply into the fabric of society, in the construction of dogs (figuratively and literally), and in praxis of keeping dogs, and this creates a caring-controlling paradox. In their book, *A Dog's World*, Jessica Pierce and Marc Bekoff (2021) discuss the hypothetical phenomenon of a post-human future—a world devoid of humans—and what that humanless future may look like for dogs. The theoretical approach of posthumanism (a critical theory in response to anthropocentrism) is gaining the interest

DOI: 10.1201/9781003361459-6

of animal studies scholars, ART scholars, and others in complementary fields. Posthumanism doesn't necessarily mean a world without humans as Pierce and Bekoff envision in their book but one that offers a new epistemology that decentres Cartesian dualism and unsettles traditional human exceptionalist paradigms and instead aims to undermine traditional species boundaries. Broadly speaking, posthumanism can be understood as a decentring of humans and a recognition that other animals are a critical part of a (post)human life. Thus, it challenges both humanism and the categorical symbolic meaning of what it is to be "human." We know that multispecies relationships are integral to many families, and separating dogs from human homes is not the intent. Rather, the post-human life shifts the paradigm to a new way of thinking and living. In fact, some scholars have suggested that the increasingly widespread practice of "pet keeping" in post-industrial societies are indicative of a shift to post-human enlightenment, which perhaps is partly true though not yet near its fruition.

It may be impossible move "beyond" violent modernist tendencies if we fail to confront the emotional investments and human conveniences that rely upon and produce dogs as a part of this caring-controlling paradox. However, is it possible to imagine a world where speciestist social conditions can be modified by disrupting the very binaries that produce human/other animal as separable? If this is the way forward, even if gained only in successive approximations (shaping into desired outcomes of some ultimate goal of [total?] liberty), then a shift to a post-humanist perspective, one that aims to correct the pernicious persuasion of human exceptionalism, should also then consider the topic of canine (and of course other species) consent and autonomy. This requires centring dogs and decentring humans, which challenges the status quo and aims to move us beyond prejudicial species oppression.

6.2   MORALITY AND ETHICS OF CARING FOR DOGS

Throughout this book, I have explored how human exceptionalism creates a culture that both loves and cares for dogs but that disempowers and disregards their unique experiences of the world. Because dogs have been created and shaped by society to render them as life-long dependents, we have an ethical duty to consider dogs as individuals who are independent from their human connections, accounting for their actual needs, not only our own. How can we envision this shift given the vast history of interconnectedness humans have with dogs? Pierce and Bekoff (2021, p. 160) write that "our [projected] journey into a post-human future is relevant not only to scientists who study dogs,"—who, they suggest, tend to look back into dogs' evolutionary past rather than at the trajectory of their future—"but also to millions of people who share their home with a canine companion. By thinking about a future when dogs go wild, we can learn a lot about dog–human relationships right now." And perhaps that projection into a post-human existence is available to some extent, if we look at the lives of free-living dogs. Though they may experience the peripheral effects of humans, their behaviour is less directly shaped by daily management.

The relationship between dogs and humans is intrinsically asymmetrical in three crucial ways: **power**, **dependency**, and **vulnerability**. Dogs are always vulnerable to the vagaries and failings of their human caregivers. Partly this is exemplified in the fact that they are unable to leave, seek help, or hold their abusive or neglectful owners accountable, but also because there is no standard of consent for dogs and consideration of nonconditional autonomy. While, yes, companion dogs are perpetually dependent on humans for the fulfilment of their basic needs when living in human homes, they are also dependent on humans to advocate for their right to bodily autonomy, which is rarely taken seriously. Power ultimately always lies in the hands of humans; they always have decisive control, even when providing dogs with any semblance of autonomy in day-to-day living. This means that dogs have no private life and are largely disempowered by limited opportunities for self-determination—they lack control over almost all aspects of their daily lives from start to end.

It is worth noting, again, the vulnerability of dogs in dog–human relationships is not incidental nor natural but a result of our desire to have animals live in our homes as our companions. Therefore, many of the vulnerabilities that dogs experience are created by us to satisfy our desires or need for convenience. I am sure a genuine relationship can develop, but it's almost always a relationship designed by humans, for humans. I absolutely do think dogs can benefit in some ways from living with humans, but it should be our duty to do so with a dog-centred approach and with one that considers their right to consent. I hope I am able to provide Juno with a good life, but I do think about the ways that I have shaped that life for her. In order to make this life as just as possible, consent needs to be taken seriously and it is my relationship with Juno and others that has inspired me to construct an objective definition that fits canine needs.

6.3   DEFINING CONSENT: CANINE-INDEXED DEFINITION

In Chapter 1 we looked at some of the human definitions about consent. Some of these are applicable to dogs while others may not be because they are written by humans and for humans (and the unique abilities humans possess). Humans experience their world in different ways than dogs and are thus not all forms of consent are germane for considering the canine perspective in their existing design. Therefore, it's time we index a definition that explicitly features dogs and their unique Umwelt. It should be a definition that encapsulates all the ways that consent does apply to dogs and how.

First, caring for dogs means that their agency is more or less a sense rather than a truth. As we have discussed throughout this entire book, dogs are dependent on humans; humans *own* dogs. Dogs have few rights within human law, and as such, we have the ultimate control over their lives, and this needs to be accounted for when considering the ways in which we can apply consent in dog–human interactions. Even when we use the least intrusive methods to teach a dog, conditioning a dog to like something (and increase the likelihood they will choose that behaviour in the future by increasing the relevance, effectiveness, and efficiency of that behaviour) is choice by design. However, we all operate on the foundation of conditioning. We all learn to function based on our knowledge of past events and skills gained through operations of classical and operant conditioning. This is the same fundamental way our dogs learn and gather information to exercise agency and make skilful decisions when the option is available to them. Not to mention, a dog's uninhibited choice may in fact be one that has been shaped by unrealistic environmental conditions or unrealistic expectations that have ultimately fostered specific behaviours, including behaviours that are not conducive to living successfully in a human-centred environment. Thus, intervention to guide them into more desirable outcomes can be crucial. If done with the aim to improve emotional wellness, behaviour modification can be the most humane and compassionate option. But we have to remember that any behaviour change we create is potentially limiting what options are available to choose from. For example, many people teach their dog to actively choose to enter their crate. But how many dogs would genuinely choose to run eagerly into a cage without some type of reward system for doing so (or punishment for not doing so)? However, the crate could be a useful endeavour for a young puppy who needs a nap or to keep them safe while unsupervised, in which case, we have to remember that we should take the least intrusive approach to affect their best outcomes, provided it isn't only in our personal best interest but to improve their experience.

Critical Corner: When making decisions on behalf of dogs in our lives we must first ask ourselves, does this benefit my dog? If yes, how? Are there alternatives to this decision, and would the outcome for them be better or worse? Ultimately, what is the best choice for them now and in the long term?

One of the things that should help us interact with dogs with better accuracy is creating an indexed definition of canine consent. This definition is based on the canine perspective (from what we can best understand about them and their behaviour). It is important to remember, however, that science is always advancing, and adjustments may need to be made as we move into a new dawning of understanding. As we dive into this definition, keep in mind that interactions with dogs should be designed as a two-way conversation, not an exact equation. It is about listening, learning from one another, and asking the right questions.

6.3.1 The Indexed Definition

The following is my definition of canine consent to be applied during dog–human interactions. It includes a general definition and five subsets of canine consent.

Canine Consent [keɪˈnaɪn kənˈsent]: a dog who gives approval, assent, or is in agreement to interact with a human. Consent is given through either learned consent behaviours (start and stop indicators) or through innate canine body language signalled through

1. Connection-seeking behaviours, signals of enjoyment where integrity is maintained through a consent test, or metacommunication signals (consent given).*
2. Stress or avoidant signals within the categories of displacement behaviours, appeasing signals, stress behaviours, and learned indicators to withdraw consent (consent not given or withdrawn).*

These signals can be found within the following contexts (subsets of canine consent):

Refer to Tables 2.1 and 2.2 *in* Chapter 2: *Description of yes and no signals*

Touch/Interaction Consent: Touching a dog should never be assumed consensual. It is important to become versed in dog body language because there are many signals, gestures, behaviours, and movements that are commonly misinterpreted. There are a few ways to navigate touch consent:

1. Allow the dogs to come to you and ask for touch.
2. Never approach a dog while they are eating, sleeping, or engaged in other activities as this is a violation of their bodily autonomy, and they do not have the proper time or information to consent.
3. Ask the dog if they want to be touched by inviting them into your space. Do they move away or remain neutral? This is a no. Do they orient their proximity, body position, and/or gaze direction toward you? This is a yes.
4. Testing consent should be done regularly (See Figure 6.1). At ~10 second intervals, stop petting and watch the dog's body language. If they do not actively seek more tactile interaction, this is a withdrawal of consent.

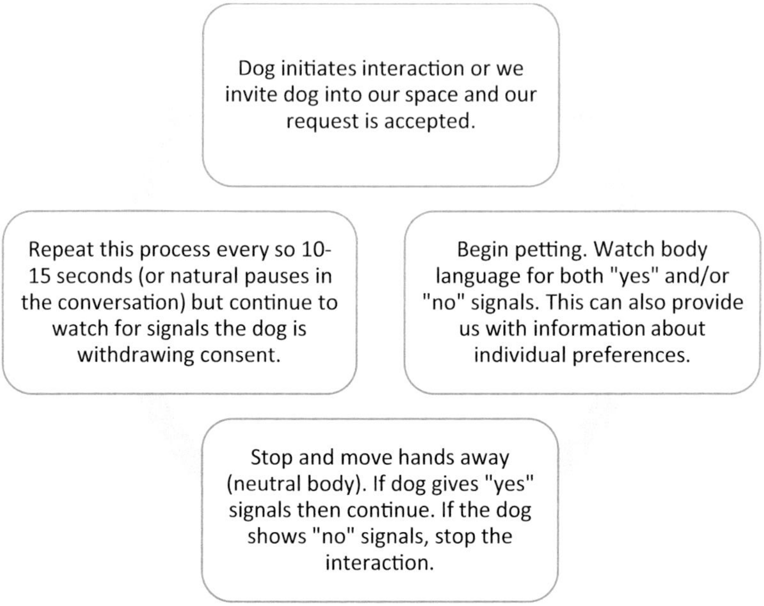

Dog initiates interaction or we invite dog into our space and our request is accepted.

Repeat this process every so 10-15 seconds (or natural pauses in the conversation) but continue to watch for signals the dog is withdrawing consent.

Begin petting. Watch body language for both "yes" and/or "no" signals. This can also provide us with information about individual preferences.

Stop and move hands away (neutral body). If dog gives "yes" signals then continue. If the dog shows "no" signals, stop the interaction.

Figure 6.1 **A consent test is a consent assessment tool. It is a protocol for asking a dog if they consent and then periodically checking if the interaction remains consensual. Think of this as a conversation, sometimes checks happen at natural pause points, and sometimes petting is obviously mutual and fewer check-ins are adequate. The more this type of conversation is practiced, the easier and more natural it becomes for both humans and dogs.**

Cooperative Care Consent: A dog should be taught at least one behaviour that indicates their consent or refusal/withdrawal of consent during various procedural handling scenarios during veterinary care or grooming. A dog who has properly learned a consent behaviour should be additionally informed with predictive cues—environmental and/or otherwise conditioned visual or verbal signals that provide the dog with information about what type of handling is about to happen. These discriminative stimulus cues provide valuable information that can help a dog make an informed decision on whether or not to consent.

Activity Consent: A dog should be allowed to choose whether to consent/withdraw consent to participate in most activities. Activities include fulfilling basic needs (e.g., daily activities that might include outings, sleeping, sniffing, play, social contact, and even learning sessions). Dogs should always be provided with the opportunity to engage with other optional activities and/or walk away from any of these activities. Mindfulness should be paid to avoid coercion—food and toys can be highly engaging and reinforcing but may also act as leverage to continue with activity engagement.

Frequent consent tests are advised for improved accuracy when gauging consent/consent withdrawal (Figure 6.1).

Consent-Based Learning: Learning is an important part of building a dog's skills to successfully navigate their environment.

1. Intentional application of punishment should always be avoided. Punishment acts as a persistent threat to behave in a particular way in order avoid potential aversive consequences. Behaviours under threat of punishment can always be assumed to be non-consensual.
2. Positive reinforcement is a great way to teach new/replacement behaviours and skillsets. However, highly reinforcing consequences can also potentially act as leverage rather than reward or reinforcement. Consent should be genuinely given, and food/attention or play should not always be contingent on performance. That way we have a strong indication that engagement in a learning session is consensual, not coercive. The aim is for the dog to choose to engage in learning because it is enjoyable and enriching.

The overall goal of teaching skills is to provide a dog with sensibilities to make informed decisions in particular social situations and reduce management over time.

Substitutive Consenting—for Decisions Made on Behalf of Dog: If there is no option for the dog to consent/withdraw consent given their dependency on humans to care for their long-term well-being, they should not be asked if they consent or not. Asking for their consent and then overriding their decision to decline or withdraw consent is damaging to the trust built within the dog–human relationship. *Not asking for consent should only be considered if there is truly no option for a dog to say "no" (example, emergent medical treatment).*

Substitutive Consent should be weighed carefully to address four key questions:

1. Does this decision best serve the well-being of the individual dog (emotionally and physically) in the bigger picture? In other words, is this a situation where you can use your knowledge of potential outcomes to make a difference to their overall and long-term well-being?
2. Are there alternatives to your decision and if so, which decision best addresses their overall and long-term well-being?
3. Are you confident that this decision is not based on your own convenience or your own best interest instead of that which best serves the individual dog?
4. How can you approach this decision in a way that is the least intrusive for the individual dog in this particular situation?

6.4 A NOTE ON DOGS AS TOOLS (RESEARCH, SERVICES, THERAPY)

You likely noticed that this book doesn't delve into a discussion about dogs used in research (medical or otherwise), dogs used for specific services or work (sight-dogs, medical or wildlife detection dogs, police dogs, customs dogs, working farm dogs, etc.), or therapy dogs (dogs used in facilitating people in clinics, hospitals, schools, etc.). I do feel that these areas are important to address, however, all of these dogs have unique circumstances and provide a specific utility to humans. Perhaps certain roles that these dogs fill (all or a few) might be problematic to the topic of consent in various ways and some more exploitative than others, though I imagine there is more variation on how exploitative a relationship might be within the various classifications than externally, and there are always going to be peripheral cases, which even extend to dogs in the role of companion.

In some interactions within each sub context, consent may apply, while in others it may not. Ultimately, companion dogs and dogs used in various forms have some analogous limitations: Training, management, confinement, and control. I hope that this book can be applied to dogs in all situations and in all environments and within all of their connections with humans. Because if consent is not an option as previously listed and interactions with humans are not in their best overall interest, we should be rethinking the position of dogs within that particular area of service.

6.5 CONCLUSION

Though the concept of consent and how this applies to dogs feels arduous and difficult to conceptualise at times, the definition I propose is important in a few ways. Specifically adjusting the definition of consent to one that applies specifically to dogs is important because it pushes the boundaries of change. Specifically using the word "consent" means that we are adjusting the definition itself to match the specific experiences that dogs have, not the other way around. Inclusion of the word "consent" is as important as the definition itself. It is humans that have to adjust their way of thinking about consent and how consent is accomplished beyond the human experience—a shift to a posthumanist consent; and a *canine consent* is just the beginning.

I hope this book opens discussions about consent for other animals in human care as well. Surely, we can construct an indexed definition of consent for other species now that we have the ball in motion. In some situations, it may be difficult for people to conceptualise, but I predict we will soon find ourselves in the midst of a paradigm shift, one that decentres ourselves and equally considers the needs and desires of both members of our relationships with other animals.

REFERENCES

John, F. J. D. (2011, January 2). *"Wonder dog's" 2,500-mile odyssey put Silverton on the map.* www.offbeatoregon.com

Pierce, J., & Bekoff, M. (2021). *A dog's world: Imagining the lives of dogs in a world without humans.* Princeton University Press.

Note: Numbers in *italics* indicate a figure and numbers in **bold** indicate a table on the corresponding page.